Introduction

0.0 the unique characteristics of the book

The main difference between this book and many other books is Chapter 3 of this book which is the application of the knowledge and skills you will be gaining through the example problems and explanations in this book along with common sense and all the tools at your disposal because in real scenarios you need all the help you can get and to apply knowledge of theories on real-life scenarios common sense is must.

In this book, I have tried to go over the concepts and theories necessary to carry out the calculations of the subject matter of the chapters, but I have also many example problems approximately more than 20 example problems per chapter. I have always felt that those like me who are ordinary and don't have exponential talent need as many examples as necessary and thus I have tried to add as many examples problems as I could to make the topics clear.

0.1 Instructions for studying the contents of the book

I have provided as many examples as I could within the time I had to write this book. But this can be a double-edged dagger because if you don't try to solve any problems without looking at the solution your chances of actually learning something will decline. So, my suggestion would be apart from the 1st 3 example problems of each chapter you should try to solve the example problems yourself self, and only when you have given it a try and found a solution then try to look at the solution provided here to see if your solution was correct or not and if you made a mistake where did you make the mistake. For those who want to tackle the 3rd chapter without any help whatsoever I'd suggest you take the self-practice one step further after you have solved a problem try to see just the answer and if it doesn't match then without looking at the solution try to redo the problem and keep a sharp eye as to what could be the mistake your making or is it that the book solution might be wrong? To human is err so you never know.

"Trial and error with a pinch of investigation is the recipe for transforming theory into practice" – Zahed Zisan

Here are the instructions in bullet points:

- **Potential Pitfall:** Avoid relying too heavily on the solutions, as it can hinder the learning process.
- **Initial Approach:** except for the first three example problems in each chapter:
  - Solve them on your own first.
  - Check the provided solutions after attempting them to verify your answers.

- **Self-Practice Strategy:**
  - Attempt to solve the remaining example problems without looking at the solutions first.

- After finding a solution, compare it with the provided solution to check for correctness and understand any mistakes.

• **Advanced Practice (in preparation for self-exploration of Chapter 3):**

- Solve the problems independently.
- Check only the final answers without looking at the detailed solutions.
- If your answer is incorrect, try to solve the problem again, identifying potential mistakes or considering the possibility of an error in the book's solution.

Links in one place:

As it would be difficult for the readers of the hardcopy readers of the book to access the additional support video links. You can find my YouTube channel as well as other resources from this one simple link, that you can easily type on your browser's search bar (or simply search: linktree Zahed Zisan). From there go to my YT channel where the playlist for this book is titled- **Applied Engineering Mechanics: Forces and Moments Book problem solutions**

Link: https://linktr.ee/zahed_zisan **(or you can just search YouTube Zahed Zisan on Google)**

0.2 Prerequisites:

The prerequisites for this book are:

1. Basic Geometry Knowledge
2. Basic knowledge of trigonometry

These topics of prerequisites are also included in a few practice problems of chapter 3. The appendix of the book will help you if you don't have the bare minimum knowledge of the prerequisites.

Contents

Introduction 1

    0.0 the unique characteristics of the book 1

    0.1 Instructions for studying the contents of the book 1

    0.2 Prerequisites: 2

Chapter-01: Concurrent Coplanar Forces 4

    1.1 Force Vector 4

    1.2 2D Force system 7

    1.3 Force resultants and components 8

    1.4 Lami's Theorem 43

Chapter-02: Moments & Couples 47

    2.1 Moment of a force 47

    2.2 Couple Moment 61

Chapter-03: Practical/Practice Problems 64

Zahed Zisan's Sample Exam 76

Appendix 78

Appendix Problems 79

Chapter-01: Concurrent Coplanar Forces

1.1 Force Vector

In structural analysis, we primarily focus on forces to understand how structures behave under different loads, which result in stress and strain. Stress is defined as force per unit area, and strain is the deformation resulting from stress. Directly measuring stress and strain can be complex. Therefore, we analyze the forces acting on an object to determine its stress-strain behavior.

This behavior depends on several factors, including the magnitude and direction of the forces, as well as the size, shape, and material properties of the object. To accurately represent and calculate these forces, we use vectors, which allow us to consider both the magnitude and direction of the forces.

The following is a force vector on a certain point:

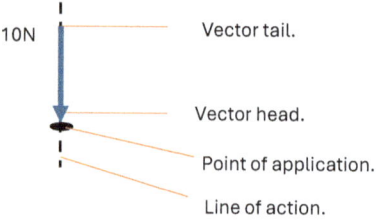

Figure-1

We characterize a vector with a tail, head, and magnitude. The magnitude can be either represented by the length of the vector or it can be written alongside it as shown in Figure-1.

Now we know that to completely define a force we need three things:

1. **Magnitude**
2. **Point of application**
3. **Direction**

Now let's understand why it is that way with the help of Newton's laws!

Let's first understand what a force with Newton's first law is. From Newton's first law, we know that a body remains at rest or moves at a constant velocity unless a force acts on it, thus a force is anything that causes a mass to change its velocity.

Now, when you sit on a chair or stand on a floor your weight which is caused by your mass and gravitational pull exerts a force on the chair or the floor, but we can not see force, but if the chair or the floor breaks due to you standing or sitting on it then we can understand the effect of the force so in other words we can only know a force has influenced as mass by observing the change in velocity of that mass.

Now let's also apply Newton's second and third laws to the example of standing on a floor or sitting on a chair.

Newton's First Law (Law of Inertia):

A body at rest will remain at rest, and a body in motion will continue in motion at a constant velocity unless acted upon by a net external force. This law describes the tendency of objects to resist changes in their state of motion.

Newton's Second Law (Law of Acceleration):

The acceleration of an object is directly proportional to the net force acting on it and inversely proportional to its mass. This can be mathematically expressed as $F = ma$. where F is the net force acting on the object, m is the mass of the object, and a is the acceleration produced.

$$F = mass \times \frac{dv}{dt}$$

with "Proportional" and "Constant" labels

Newton's Third Law (Action and Reaction):

For every action, there is an equal and opposite reaction. This means that if object A exerts a force on object B, then object B exerts an equal and opposite force on object A.

Now if we focus on the example scenario except instead of yourself consider a ball:

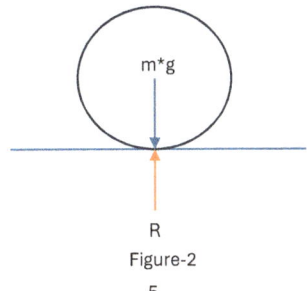

Figure-2

The ball's mass when multiplied by the effects of gravity becomes a weight that acts on the floor and as the 3rd law implies the floor gives a reaction, now as we have discussed with the first law if there is no change in velocity then there will be no visible or accountable effect of the force thus the force has a **Magnitude** which when greater then the reaction it effects would be visible to our human eyes in a way we can perceive its effects. Now if the object wasn't a ball that was only one point of contact but an aquarium:

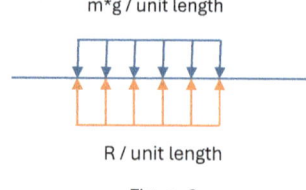

Figure-3

Then the total weight, instead of being applied on a single point the weight of the aquarium would have been spread out throughout the contact area thus the reaction would have been different (spread out). Indicating the effect of the **Point of application.**

Now if there was another force maybe you kicked the ball then there would be three forces in that scenario, the weight of the ball the reaction of the floor, and your kick force but then which direction would the ball go? If the weight is greater, then the kick force and the floor's reaction then the floor would break, and the ball would go down but if the kick were greater in force, then? To answer this, you'd ask in which **Direction** did I kick the ball? To figure out the direction you also need the point of application.

So, from this, I hope we are clear about the 3 requirements or characteristics of a force or force vector.

1.2 2D Force system

When some forces (2 or more) act on a body we call it a force system.

If 2 or more forces are acting on the same plane, then they are called a system of coplanar forces.

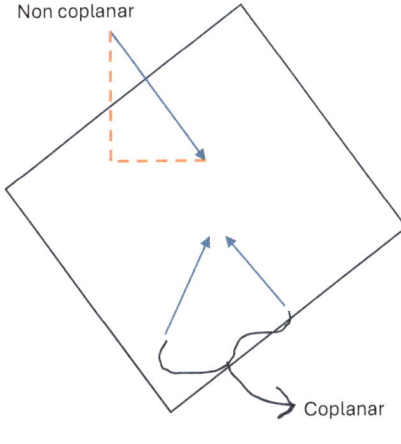

Figure-4

If the lines of action of 2 or more forces intersect the same point, then the forces are concurrent.

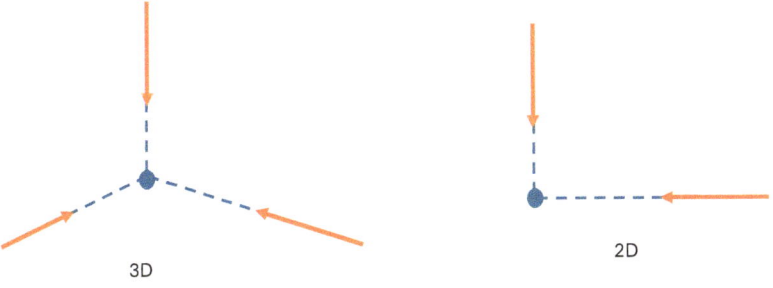

Figure-5

So, if 2 or more forces work on the same plane and their lines of action intersect at the same point then this system of forces is called concurrent coplanar forces.

1.3 Force resultants and components

Any force vector can be broken down into two or more component vectors. Conversely, we can determine a resultant force vector by combining two or more component vectors. In 2D force systems, it is helpful to consider the orthogonal (right-angled) components of a force. Two vector components at right angles to each other can combine to produce the same effect as the original force when applied together.

Example-1: Calculate the components of the force vector shown below.

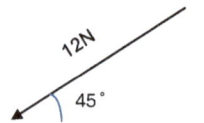

Solution:

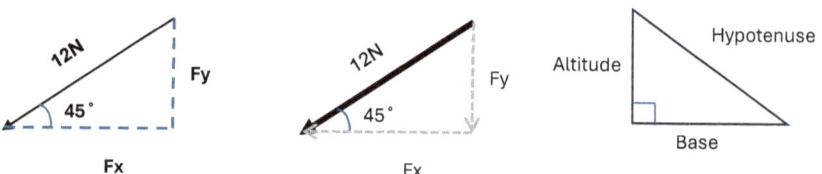

Fx = 12*Cos (45) = 8.49N   [$Cos\theta = \frac{base}{hypotenous}$ thus $Cos(45) = \frac{Fx}{12}$ thus Fx = 12Cos(45)]

Fy = 12*Sin (45) = 8.49N   [$Sin\theta = \frac{altitude}{hypotenous}$ thus $Sin(45) = \frac{Fy}{12}$ thus Fy = 12Sin(45)]

Before you read or move forward to the next part, if you are having trouble understanding the calculations of the above example problem then it would be a good idea to revisit simple geometry and trigonometry and try to do the appendix problems to get a better handle of this kind of calculations. Otherwise, just move to the next part continuing with the black color text.

Now from Example-1, we can come up with a shortcut that will make our lives easier. When the angle known is with the Y direction force we will use Cosθ and when the angle is not with the Y direction we will use Sinθ and vice versa with X direction force. You will feel what I am talking about when you do the next problem sets. Before the next problem, we will learn about sign convention and 2D coordinate systems.

2D Coordinate System

Imagine a big piece of paper with a grid on it. This grid helps us find the location of points and the directions of arrows (called vectors). The grid has two main lines called axes:

X-Axis: This is a horizontal line (goes left and right).

Y-Axis: This is a vertical line (goes up and down).

Where these two lines cross is called the origin. We usually label this point as (0,0).

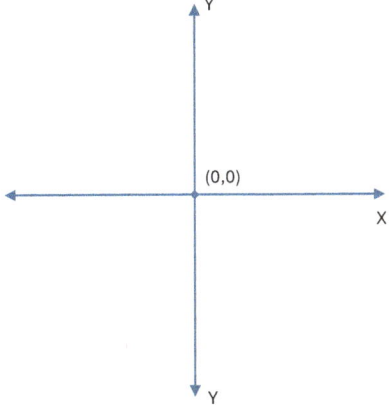

Figure-6

Quadrants

The axes divide the paper into four parts called quadrants:

First Quadrant: Top-right (both x and y are positive).

Second Quadrant: Top-left (x is negative, y is positive).

Third Quadrant: Bottom-left (both x and y are negative).

Fourth Quadrant: Bottom-right (x is positive, y is negative).

Points on the Grid

A point on the grid is like an address for a location. It is written as (x, y):

x: Tells you how far to move left or right from the origin.

y: Tells you how far to move up or down from the origin.

For example, the point (3, 2) means move 3 spaces to the right and 2 spaces up.

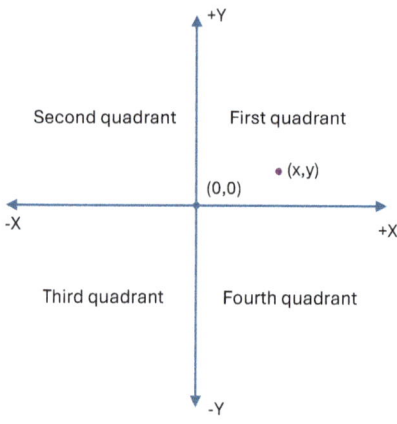

Figure-7

Sign Convention

When we talk about directions, we use signs (+ or -) to show which way we're moving:

Positive x (+x): Move to the right.

Negative x (-x): Move to the left.

Positive y (+y): Move up.

Negative y (-y): Move down.

Force Vectors

Forces can be shown as arrows on this grid. Each force has:

A magnitude (how strong it is, like 10 N).

A direction (which way it's pushing or pulling).

Breaking Down Forces

When a force is not exactly along an axis, we break it into two parts:

One part is along the x-axis (horizontal).

One part is along the y-axis (vertical).

This helps us understand how the force moves things in each direction.

Example-2: Calculate the components of the force vector shown below.

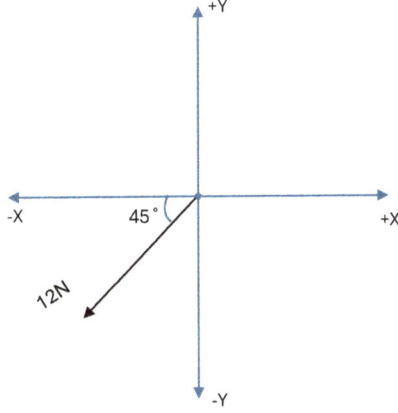

Solution: if you draw some dotted lines to imagine a triangle then this problem is no different from the problem of Example-1

The force is moving downward and to the left, now recall the sign conventions stated previously.

Fx = -12*Cos (45) = -8.49N (negative meaning leftward)

Fy = -12*Sin (45) = -8.49N (negative meaning downward)

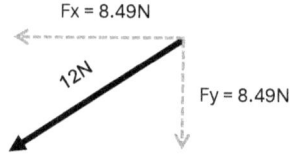

But there is no need to imagine the triangle as we have discussed we can simply do the following to make it easier and with the following examples, this will become clearer.

When the angle known is with the Y direction force, we will use Cosθ and when the angle is not with the Y direction, we will use Sinθ and vice versa with X direction force.

In Example-2 we calculated the components of a force, similarly, we can do the opposite and calculate the resultant of component forces. Before moving on to Example-3 let's take a look at a few concepts with an example (even though I am going to say imagine, try to draw up the scenario using pen and paper or any other means, which now should be quite easy for you after the 1st two examples).

Resultant Force

Imagine you are pushing a heavy box. Now, suppose your friend also pushes the same box but from a different direction. The box will move in a new direction due to the combined effect of both pushes. This combined effect is called the resultant force. It's like finding the overall push or pull on an object when multiple forces act on it at the same time.

How to Find the Resultant Force

To find the resultant force, we need to add up all the individual forces acting on the object. This can be done by breaking down each force into its x (horizontal) and y (vertical) components and then combining these components.

Steps to Calculate the Resultant Force

1. Resolve Each Force into Components:

   - If you have a force F making an angle θ with the horizontal or vertical(whichever would be easier for analysis), you can find:

     - The x-component: $Fx = F \cos(\theta)$

     - The y-component: $Fy = F \sin(\theta)$

2. Sum the Components:

   - Add up all the x-components to get the total x-component (Rx).

- Add up all the y-components to get the total y-component (Ry).

3. Calculate the Magnitude of the Resultant Force:
   - Use the Pythagorean theorem: $R = \sqrt{R_x^2 + R_y^2}$
   - This gives you the overall strength of the resultant force.

Example Calculation

Imagine two forces acting on an object:
- F1 = 4 N to the right (along the x-axis).
- F2 = 3 N upwards (along the y-axis).

Here's how you find the resultant:

1. Resolve the Forces:
   - F1 has components F1x = 4 N, F1y = 0 N.
   - F2 has components F2x = 0 N, F2y = 3 N.

2. Sum the Components:
   - Total x-component: Rx = F1x + F2x = 4 + 0 = 4 N
   - Total y-component: Ry = F1y + F2y = 0 + 3 = 3 N

3. Calculate the Magnitude:
   - $R = \sqrt{R_x^2 + R_y^2} = \sqrt{4^2 + 3^2} = \sqrt{16 + 9} = \sqrt{25} = 5$ N

Angle of the Resultant Force

The angle of the resultant force (θ) tells us the direction of the resultant force relative to the horizontal (x-axis) or to the vertical (y-axis). It can be calculated using the inverse tangent function (also known as arctan or $\tan^{-1}$).

Steps to Calculate the Angle

1. Find the Angle:
   - Use the formula: $\theta = \tan^{-1}(R_y / R_x)$ [for angle with the horizontal]
   - Use the formula: $\theta = \tan^{-1}(R_x / R_y)$ [for angle with the verticle]

- This gives the angle in degrees or radians, depending on your calculator settings.

Example (continued)

Using the previous example:

- $Rx = 4\ N$
- $Ry = 3\ N$

Calculate the angle:

- $\theta = \tan^{-1}(3/4) = \tan^{-1}(0.75) \approx 36.87°$

So, the resultant force of 5 N makes an angle of approximately 36.87 degrees with the horizontal.

Instead of Rx and Ry like the above explanation we will sometimes be using the terms ΣFx (Rx) and ΣFy. If you are having a hard time with the above example, don't worry too much about it as we will see more example problems below with figures which will help you understand better.

Example-3: Calculate the resultant and angle of application of the following concurrent coplanar forces.

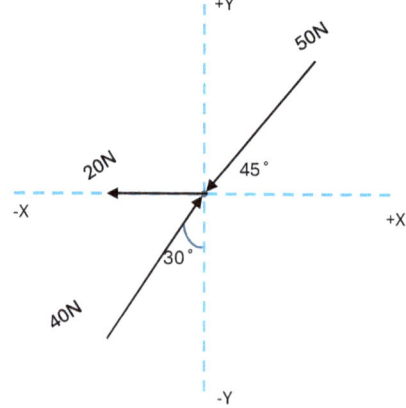

Solution:

$\Sigma Fx = -20 - 50\cos 45° + 40\sin 30° = -35.36N$

$\Sigma Fy = -50\sin 45° + 40\cos 30° = -0.714N$

$R = \sqrt{\Sigma Fx^2 + \Sigma Fy^2} = 35.37N$

Now for the angle of application, we will determine the angle with the vertical axis you can choose to determine the angle with the horizontal first if you wish to.

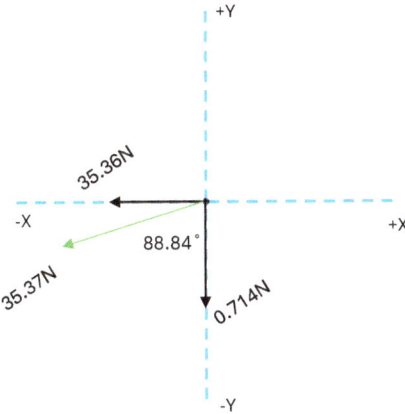

$\theta = \tan^{-1}(35.36 / 0.714) = 88.84°$ with the vertical

thus $90 - 88.84 = 1.16°$ with the horizontal $[\theta = \tan^{-1}(0.714 / 35.36) = 1.16°]$

Now before we move on to the next problem we will learn about the Global and Local axis, the difference between them to be precise.

Global and Local Axis

The global axis is the universal axis that we have been using so far. The sign convention and its setup are shown in Figure 7.

The local axis on the other hand is an axis system that is free from the universal axis restrictions. Sometimes to make an analysis easier or to simplify complex scenarios local axis is used instead of the global axis and once the required results are found in the local axis those can be converted to the global axis for the sake of communicating the results easily with the universally known global axis system to anyone. Examples of a few local axes are shown below in Figure 8.

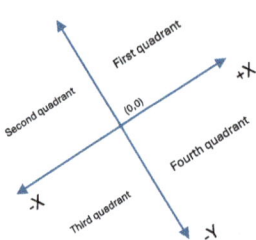 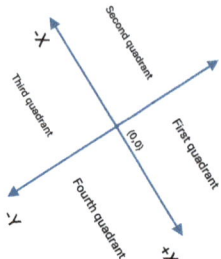

Figure-8

Understanding Local and Global Axis Systems

Imagine you are playing with a toy car on a big table. The table represents the entire world, and the car has its own little world. To keep track of where everything is, we use something called an "axis system."

Global Axis System

The global axis system is like the map of the entire table. It helps us understand where things are in the whole room. This system usually has:

- X-axis (left-right)

- Y-axis (forward-backward)

- Z-axis (up-down)

Think of this as a big grid laid out over the table. No matter where you place the toy car on the table, you can use this grid to find its exact position.

Local Axis System

The local axis system is like the map for the toy car itself. Imagine if the car had a little grid on its roof:

- X-axis (left-right for the car)

- Y-axis (forward-backward for the car)

- Z-axis (up-down for the car)

This helps to understand where things are in relation to the car. For example, if you say "move forward," the car knows to move in the direction it is facing, not based on the entire table.

Transforming Between Global and Local Axes

To transform between these systems, we need to understand where the car is on the table and how it is oriented.

From Global to Local

1. Identify Position: Find the car's position on the global grid (table).

2. Consider Orientation: Determine which way the car is facing (its rotation).

3. Translate Coordinates: Adjust the global coordinates based on the car's position and rotation.

For example, if the car is at (5, 5) on the table and facing forward, you can say that the front of the car is 1 unit forward from the car's perspective.

From Local to Global

1. Identify Position: Start with the car's position and orientation.

2. Translate Movement: Convert the local movement (e.g., forward for the car) into the global grid.

For example, if the car moves 1 unit forward, you need to add this movement to its current global position based on its orientation.

Simple Example

Imagine the car is at position (5, 5) on the table (global coordinates) and is facing towards the top of the table.

- Global Axis: (5, 5)

- Local Axis: (0, 0) (because we're starting at the car)

If the car moves 2 units forward:

- Local Axis Move: (0, 2)

- Global Axis Move: The car moves from (5, 5) to (5, 7) in global coordinates.

If the car turns to the right (90 degrees):

- Local Axis: Forward is now to the right of the table (global X-axis increases).

- Global Axis Move: Moving forward 2 units means moving from (5, 7) to (7, 7) on the table.

This process involves some math, like adding and rotating vectors, but this is the basic idea. By knowing the position and orientation, we can easily switch between the global and local axis systems.

Example-4: Determine the summation of forces in the local X and the local Y direction.

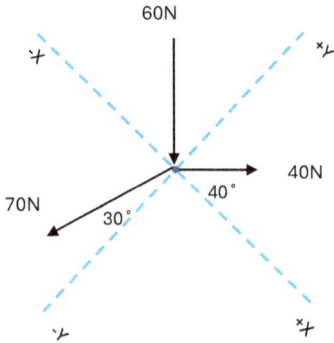

Solution:

You can see that there is an angle missing for us to carry out our calculations. And on top of that, this could be the first local axis problem you are seeing. Fear not it's very easy once you grasp that it's the same as any global axis problem just a bit rotated.

Now look at the figure below while you read this explanation. The 40N force seems to be global horizontal and the 60N force is global vertical. The 40N force makes a 40° angle with the +X axis, thus if we draw a horizontal line (shown in green) through the center then that makes a 40° angle with the -X axis (reciprocal angle concept used). Now this green line makes a 90° angle with the 60N force, thus 90-40 = 50°. The 60N force makes a 50° angle with the -X axis. This is all the angles we need to carry out the calculation. But we will determine another angle for fun. The 70N force makes a 30° angle with the -Y, and the green horizontal line we drew makes a 40° angle with the -X. The -X and -Y are 90° with each other thus 90-40-30 = 20°.

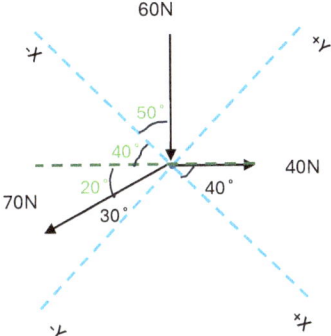

$\sum Fx = 60\cos(50) + 40\cos(40) - 70\sin(30) = 34.20N$

$\sum Fy = -60\sin(50) + 40\sin(40) - 70\cos(30) = -80.87N$

Example-5: Calculate the resultant and the angle it makes with the horizontal.

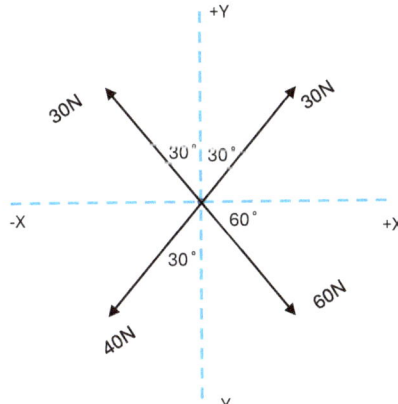

Solution:

$\Sigma Fx = 30\text{Sin}(30) - 30\text{Sin}(30) + 60\text{Cos}(60) - 40\text{Sin}(30) = 10\text{N}$

$\Sigma Fy = 30\text{Cos}(30) + 30\text{Cos}(30) - 60\text{Sin}(60) - 40\text{Cos}(30) = -34.64\text{N}$

$R = \sqrt{\Sigma Fx^2 + \Sigma Fy^2} = 36.05\text{N}$

$\theta = \tan^{-1}(34.64/10) = 73.89°$

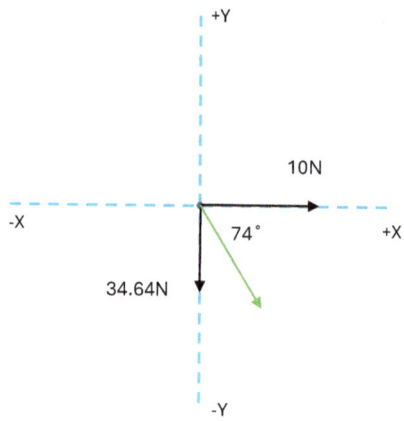

Example-6: If all the forces in the example-5 problem were 30N then what would be the resultant and the angle with the horizontal? Can you answer without calculating?

Solution:

The solution to example 6 is given after example 7 but before you are tempted to look at the example solution, I want you to think while looking at the problem very carefully. At this point, you should be able to guess the answer without calculating. Don't worry if you need to calculate to get the answer. There are plenty more problems after this that will increase your skill in this thing, and you will eventually be able to do more than this in your mind without calculating by hand or paper but to reach that level there is no alternative to practice-think-practice and repeat.

Example-7: Calculate the resultant and the angle of the resultant with the horizontal and can you guess if the resultant would be closer to the Y or the X axis before you do the calculations?

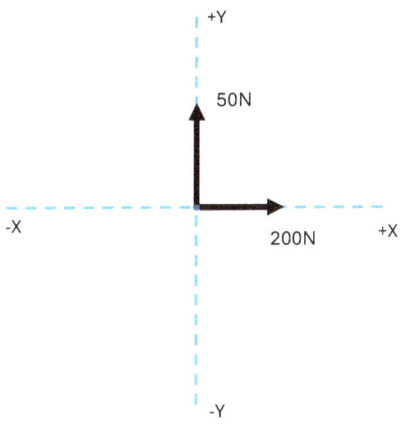

Solution:

$R = \sqrt{\Sigma Fx^2 + \Sigma Fy^2} = 206.16N$

$\theta = \tan^{-1}(50/200) = 14.036°$

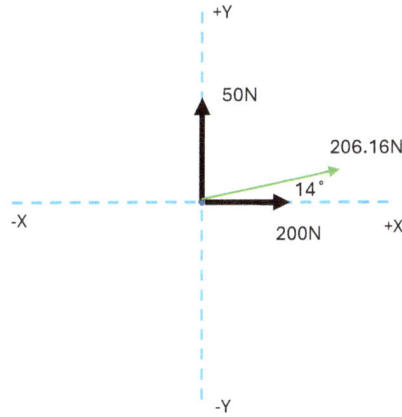

From the previous example as well as this one we can see that the resultant leans towards the larger component.

This is very easy to understand. But understanding this can lead to applications. Once you understand it you can use it to your advantage. An example of the application of this example is given in the 3rd chapter practice problem number 4.

Solution to example-6: the answer is zero. 0. Opposite same-magnitude forces cancel each other out.

Example-8: Calculate and show the direction of the components of the following force.

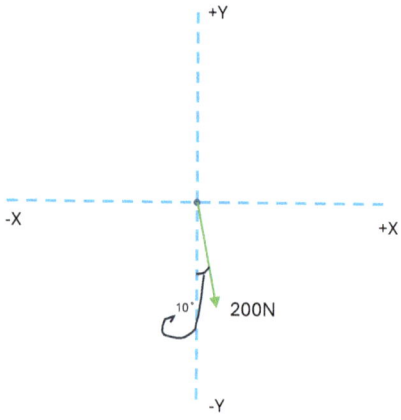

Solution:

$\Sigma Fx = 200\text{Sin}(10) = 34.73\text{N}$

$\Sigma Fy = -200\text{Cos}(10) = -197\text{N}$

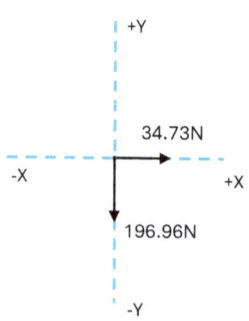

Example-9: Calculate the resultant and the angle it makes with the horizontal.

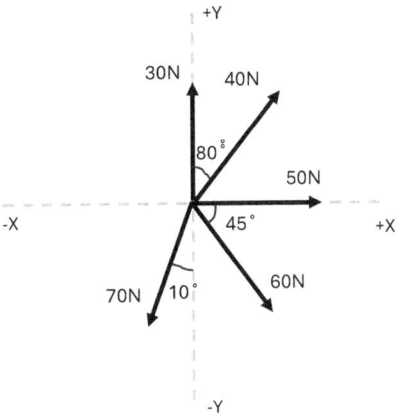

Solution:

$\Sigma Fx = 50 + 40\sin(80) + 60\cos(45) - 70\sin(10) = 119.66N$ (towards the right)

$\Sigma Fy = 30 + 40\cos(80) - 60\sin(45) - 70\sin(10) = -74.42N$ (downwards)

$R = \sqrt{\Sigma Fx^2 + \Sigma Fy^2} = 141.20N$

$\theta = \tan^{-1}(50/200) = 31.87°$

Example-10: Calculate the resultant force and the angle it makes with the vertical. The red dot is the origin point of the forces.

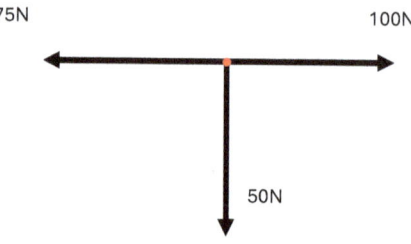

Solution:

Don't be afraid that there is no axis given. If there is no axis then just place it where ever would be simpler for your required calculation. You can also decide if you wish to use a global or a local axis.

Here we will place the global axis origin at the origin of the force, which will make our calculations easier, also because no other data is found here (like any particular distance from the origin of the forces or anything)

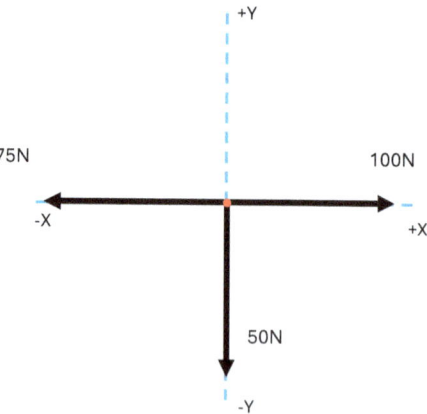

$\sum Fx = 100N - 75N = 25N$ (towards the right)

$\sum Fy = -50N$ (downwards)

$R = \sqrt{\sum Fx^2 + \sum Fy^2} = 55.90N$

$\theta = \tan^{-1}(25/50) = 26.58°$

Now before we move on to another example problem, let's talk a bit about the principle of transmissibility.

Principle of Transmissibility: The Principle of Transmissibility states that a force acting on a body can be applied anywhere along the force's line of action without changing its effect on the body.

For example, think of pushing a toy car forward or pulling it forward. Both cases give you the same result. So, if you know that a car is moving forward, and you also know that a force is the reason behind this movement then you can assume push or pull (whichever is easier for your work) in the line of movement.

Figure-9

In Figure-9, A vector represents a force of 50N's point of application and a line of action. Now according to the principle of transmissibility, the force acting on the car can be applied anywhere along the force's line of action without changing its effect on the body. Thus, the 50N force can be applied at point B.

You actually have come across this without realizing or realizing in example problem 3, where this principle could have been applied. Not that it would make a significant difference, rather it would have increased the amount of effort we needed to solve that problem. But just to get a feel of this principle we will revisit that problem now.

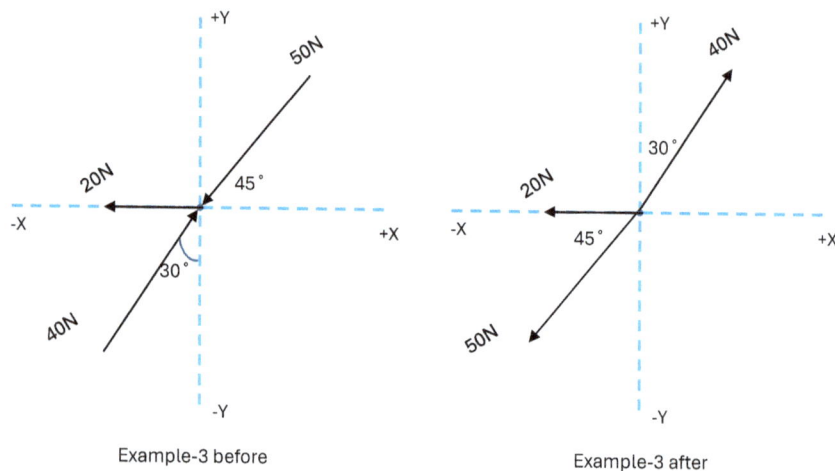

Example-3 before   Example-3 after

A little change in point like the car example and the application of reciprocal angles and even though it might look different to some it is the exact same problem with the exact same solution.

Now we will go into 2 different categories of forces, 1st Applied & Nonapplied, and 2nd External & Internal.

Applied & Nonapplied:

Applied Force

Definition: An applied force is any force that is intentionally exerted on a structure from an external source. This force comes from something acting on the structure, like wind, people, vehicles, or equipment.

Example:

- **Wind Load**: Imagine a tall building in a windy city. The wind blowing against the building creates an applied force. This force can push against the sides of the building, causing it to sway slightly. Engineers need to design the building so it can withstand these wind forces without being damaged.

Nonapplied Force (or Self-Weight)

Definition: A nonapplied force is a force that is inherent to the structure itself. It is not caused by external factors but by the weight of the materials that make up the structure.

Example:

- **Self-Weight**: Think of a bridge made of concrete and steel. The self-weight of the bridge is the force due to its own weight. This includes all the concrete, steel, and other materials used to construct the bridge. Even without any cars or people on it, the bridge has to support its own weight. Engineers must consider this force when designing the bridge to ensure it doesn't collapse under its own weight.

Simple Analogy

Imagine you have a bookshelf:

- **Applied Force**: This would be like putting books on the shelf. The weight of the books is an external force that the shelf needs to support.
- **Nonapplied Force**: This would be the weight of the shelf itself. Even without any books, the shelf has its own weight that it needs to support.

External & Internal:

External Forces

Definition: External forces are forces that act on a structure from outside sources. These forces come from things like wind, earthquakes, vehicles, and people.

Example:

- **Wind Load**: Imagine a skyscraper. The wind blowing against the sides of the skyscraper creates an external force. This force acts on the outside of the building, pushing against it and causing it to sway slightly. Engineers must design the building to withstand these wind forces.

Internal Forces

Definition: Internal forces are forces that act within the structure itself, created as a response to external forces. These forces include tension, compression, shear, and torsion that occur within the materials of the structure.

Example:

- **Tension in a Bridge Cable**: Think of a suspension bridge. The weight of the cars and trucks traveling across the bridge creates external forces that push down on the bridge deck.

In response, the cables holding up the bridge experience internal forces. These internal forces are tension forces, as the cables are being stretched to hold up the weight of the bridge deck and the vehicles on it.

Simple Analogy

Imagine you have a rubber band:

- **External Force**: When you pull on the rubber band from both ends, the force you apply is an external force. You're applying this force from outside the rubber band.
- **Internal Force**: As you pull on the rubber band, it stretches. The stretching force within the rubber band that resists your pull is an internal force. This force is what keeps the rubber band together and prevents it from breaking.

Don't worry if you don't fully understand these force categories, we will need them for things that will come later. However, I wanted you to know a little bit ahead of time, so things become easier later and also to help you put on your thinking caps.

Now let's get back to doing example problems.

Exampe-11: Find the equilibrium force for a system where forces of 250N, 320N, and 400N act at angles of 30°, 180°, 270°.

Solution:

As you can see sometimes no one will provide you with a schematic of a problem you are required to solve you will just have information, and sometimes even information won't be given, and you have to figure out that information yourselves using your skills (you will encounter this kind of problems in chapter 3). Now, to draw the figure ourselves we need to have the basic knowledge of trigonometry, which I have mentioned before earlier in this chapter.

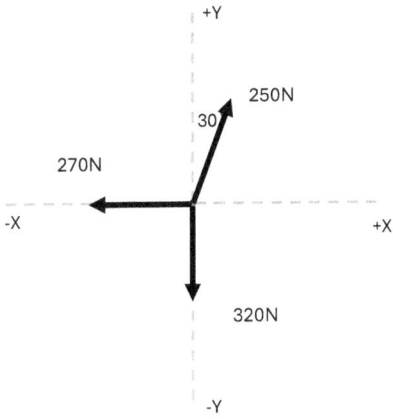

$\sum Fx = -270 + 250\sin(30) = -145N$ (leftward)

$\sum Fy = -320 + 259\cos(30) = -103.49N$ (Downward)

$R = \sqrt{\sum Fx^2 + \sum Fy^2} = 178.14N$

$\theta = 35.51°$ (the question didn't ask for this though)

Now with the next example, we will try to put our skills developed so far to solve a problem. Remember a little common sense is always required along with skills to solve a problem.

Example-12:

How much force in which direction is required to eliminate the effect of the system of forces shown below?

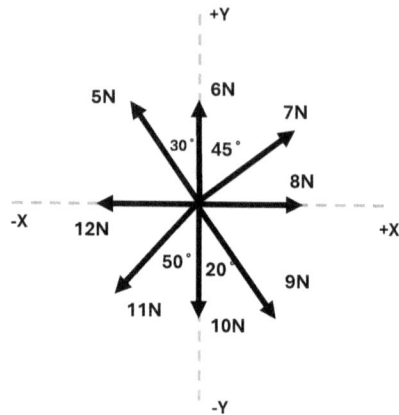

Solution:

When there are multiple forces working on a point then the effect would be the resultant and if there is movement it will be in the direction of the resultant. So, to nullify the effect of the forces it is sufficient to nullify the resultant force and we can do that by introducing a new force in the system that will have the same magnitude as the resultant and the opposite direction to the resultant. Think of it as 2 people pulling on a rope with the same force, opposite to each other, and thus the rope remains stationary as the forces cancel each other out.

You have actually seen this before in example 6.

$\Sigma Fx = 8 + 7\operatorname{Sin}(45) + 9\operatorname{Sin}(20) - 11\operatorname{Sin}(50) - 12 - 5\operatorname{Sin}(30) = -6.99N$

$\Sigma Fy = 6 + 7\operatorname{Cos}(45) - 9\operatorname{Cos}(20) - 10 - 11\operatorname{Cos}(50) + 5\operatorname{Cos}(30) = -10.25N$

$R = \sqrt{\Sigma Fx^2 + \Sigma Fy^2} = 12.41N$

$\theta = \tan^{-1}(25/50) = 56°$ with the horizontal

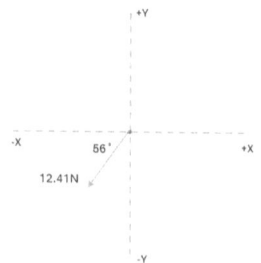

Checks are something that engineers always do. Because if their calculations are not giving the result they thought they would then there be severe consequences.

So, now we will see if what we hypothesized is actually correct by introducing the following force in red color to the system. If this addition makes the system neutral, then the check is ok.

Check:

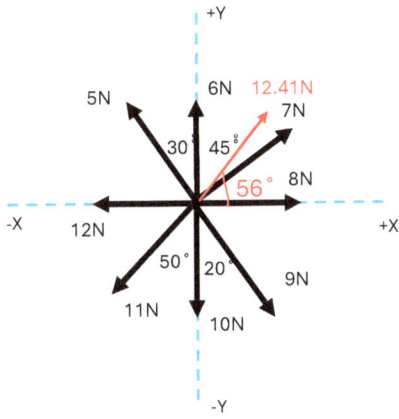

$\sum Fx = 8 + 7\sin(45) + 9\sin(20) - 11\sin(50) - 12 - 5\sin(30) + 12.41\cos(56) = 0.004N \approx 0$ N (near zero due to rounding)

Similarly,

$\sum Fy = 0N$ (do it yourself to check)

Thus, we have come up with a solution to neutralize the effect of many forces working on a point. Now, this point could be a pin/ joint of a truss or a pin where 7 cables were tied and adding the 8th cable with the right amount of tension force in the right direction to stabilize the pin.

This means you have taken your skills from the example problems and taken your 1st step towards solving engineering problems.

Before moving on to more example problems we will do a little test to see if we remember the basic concepts that we were introduced to at the beginning of the chapter.

The ropes in the following figure exert a force of 15N each. Can this be considered a concurrent coplanar force?

Figure-10

Here is a little recall for you if you are having a hard time answering this question. Figure-11 shows the line of action of the ropes in a dotted orange.

Figure-11

As you can see the lines of action are intersecting each other. And if you recall the definitions from section 1.2 then yes it is a concurrent coplanar force system.

Now let's jump to the next example. To solve this next example, you need to go back to the very first page of this chapter as this one is a graphical problem. Although it doesn't have that much fame these days, however, knowing and practicing a little bit of it will take an additional toll on your mental tool kit for problem-solving.

Example-13:

Calculate the resultant force.

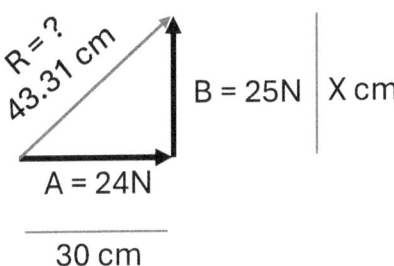

Solution:

If 24N = 30cm then,

1N = 30/24 =1.25cm.

thus 25N = 25*1.25 = 31.25cm

Thus R = 43.31cm = 43.31/1.25 = 34.66N

Example-14:

You are a trainee of an engineering team that is investigating a design done by an engineer about 100 years ago. And the engineer did graphical calculations. Your task as a trainee on their first day is to figure out the unit of force per unit length.

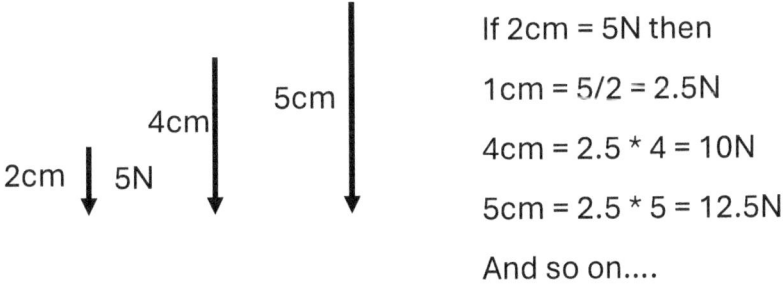

Example-15:

Determine the components of the force.

So far, we have been working with angles but what if you don't have any fancy equipment to measure angles and can only use a measuring tape or something to get the length? This example will show you how to do it in those cases. Remember once again the basics of trigonometry are required here.

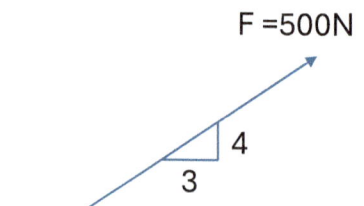

Solution:

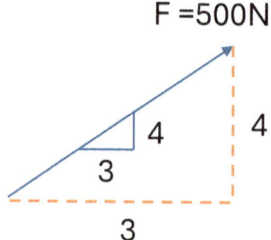 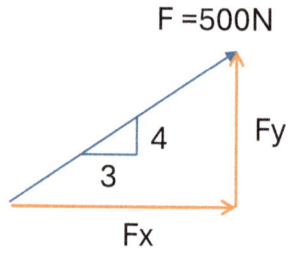

The length of the hypotenuse = $\sqrt{3^2 + 4^2} = 5$

Thus $\frac{4}{Fy} = \frac{5}{F}$ thus, Fy = 4F/5 = 4*500/5 = 400N

Similarly,

$\frac{3}{Fx} = \frac{5}{F}$ thus, Fx = 3F/5 = 3*500/5 = 300N

The next few example problems are for the purpose of more practice. There is no graphical representation like example 11 for these next problems so, you should try to draw them first and then try to solve them. The solution is also provided as these are example problems, but the solution is given in a different way than you have seen so far. So, it would be best to practice them the way we did so far. And just for insurance, you can check the solutions to see if your results match or not.

Example-16:

Find the equilibrium force for a system where forces of 200N, 150N, and 300N act at angles of 45°, 120°, and 240°.

Solution:

Resolving each force into its horizontal (x) and vertical (y) components:

- Force at 45°:
  - $F\_\{1x\} = 200 \sin(45°) = 200 * 0.707 = 141.4N$
  - $F\_\{1y\} = 200 \cos(45°) = 200 * 0.707 = 141.4N$

- Force at 120°:
  - $F\_\{2x\} = 150 \cos(120°) = 150 * (-0.5) = -75N$
  - $F\_\{2y\} = 150 \sin(120°) = 150 * 0.866 = 129.9N$

- Force at 240°:
  - $F\_\{3x\} = 300 \cos(240°) = 300 * (-0.5) = -150N$
  - $F\_\{3y\} = 300 \sin(240°) = 300 * (-0.866) = -259.8N$

$\Sigma Fx = 141.4 + (-75) + (-150) = -83.6N$

$\Sigma Fy = 141.4 + 129.9 + (-259.8) = 11.5N$

$R = \sqrt{((\Sigma Fx)^2 + (\Sigma Fy)^2)} = \sqrt{((-83.6)^2 + (11.5)^2)} \approx 84.38N$

Example-17:

Find the equilibrium force for a system where forces of 100N, 250N, and 350N act at angles of 60°, 135°, and 315°.

Solution:

Resolving each force into its horizontal (x) and vertical (y) components:

- Force at 60°:
  - $F_{1x} = 100 \sin(60°) = 100 * 0.866 = 86.6N$
  - $F_{1y} = 100 \cos(60°) = 100 * 0.5 = 50N$

- Force at 135°:
  - $F_{2x} = 250 \cos(135°) = 250 * (-0.707) = -176.8N$
  - $F_{2y} = 250 \sin(135°) = 250 * 0.707 = 176.8N$

- Force at 315°:
  - $F_{3x} = 350 \cos(315°) = 350 * 0.707 = 247.5N$
  - $F_{3y} = 350 \sin(315°) = 350 * (-0.707) = -247.5N$

Sum the horizontal components (ΣFx):
$\Sigma Fx = 86.6 + (-176.8) + 247.5 = 157.3N$

Sum the vertical components (ΣFy):
$\Sigma Fy = 50 + 176.8 + (-247.5) = -20.7N$

Calculate the resultant force (R):
$R = \sqrt{(\Sigma Fx)^2 + (\Sigma Fy)^2} = \sqrt{(157.3)^2 + (-20.7)^2} \approx 158.7N$

Example-18:

Find the equilibrium force for a system where forces of 500N, 200N, and 450N act at angles of 0°, 90°, and 210°.

Solution:

Resolve each force into its horizontal (x) and vertical (y) components:

- Force at 0°:
  - $F_{1x} = 500 \cos(0°) = 500 * 1 = 500N$
  - $F_{1y} = 500 \sin(0°) = 500 * 0 = 0N$

- Force at 90°:
  - $F_{2x} = 200 \cos(90°) = 200 * 0 = 0N$
  - $F_{2y} = 200 \sin(90°) = 200 * 1 = 200N$

- Force at 210°:
  - $F_{3x} = 450 \cos(210°) = 450 * (-0.866) = -389.7N$
  - $F_{3y} = 450 \sin(210°) = 450 * (-0.5) = -225N$

Sum the horizontal components (ΣFx):
$\Sigma Fx = 500 + 0 + (-389.7) = 110.3N$

Sum the vertical components (ΣFy):
$\Sigma Fy = 0 + 200 + (-225) = -25N$

Calculate the resultant force (R):
$R = \sqrt{(\Sigma Fx)^2 + (\Sigma Fy)^2} = \sqrt{(110.3)^2 + (-25)^2} \approx 113N$

Example-18:

Find the equilibrium force for a system where forces of 300N, 400N, and 500N act at angles of 30°, 120°, and 240°.

Solution:

Resolving each force into its horizontal (x) and vertical (y) components:

- Force at 30°: - $F\_\{1x\} = 300 \cos(30°) = 259.8N$ - $F\_\{1y\} = 300 \sin(30°) = 150.0N$
- Force at 120°: - $F\_\{2x\} = 400 \cos(120°) = -200.0N$ - $F\_\{2y\} = 400 \sin(120°) = 346.4N$
- Force at 240°: - $F\_\{3x\} = 500 \cos(240°) = -250.0N$ - $F\_\{3y\} = 500 \sin(240°) = -433.0N$

Sum the horizontal components (ΣFx): $\Sigma Fx = 259.8 + -200.0 + -250.0 = -190.2N$

Sum the vertical components (ΣFy): $\Sigma Fy = 150.0 + 346.4 + -433.0 = 63.4N$

Calculate the resultant force (R): $R = \sqrt{(\Sigma Fx)^2 + (\Sigma Fy)^2} = \sqrt{(-190.2)^2 + (63.4)^2} \approx 200.5N$

Example-19:

Find the equilibrium force for a system where forces of 250N, 350N, and 450N act at angles of 60°, 150°, and 270°.

Solution:

Resolve each force into its horizontal (x) and vertical (y) components:

- Force at 60°: - F_{1x} = 250 cos(60°) = 216.5N - F_{1y} = 250 sin(60°) = 125.0N
- Force at 150°: - F_{2x} = 350 cos(150°) = -303.1N - F_{2y} = 350 sin(150°) = 175.0N
- Force at 270°: - F_{3x} = 450 cos(270°) = 0.0N - F_{3y} = 450 sin(270°) = -450.0N

Sum the horizontal components (ΣFx): ΣFx = 216.5 + -303.1 + 0.0 = -86.6N

Sum the vertical components (ΣFy): ΣFy = 125.0 + 175.0 + -450.0 = -150.0N

Calculate the resultant force (R): R = √((ΣFx)^2 + (ΣFy)^2) = √((-86.6)^2 + (-150.0)^2) ≈ 173.2N

Example-20:
Find the equilibrium force for a system where forces of 100N, 200N, and 300N act at angles of 45°, 135°, and 225°.

Solution:

Resolve each force into its horizontal (x) and vertical (y) components:

- Force at 45°: - F_{1x} = 100 cos(45°) = 70.7N - F_{1y} = 100 sin(45°) = 70.7N
- Force at 135°: - F_{2x} = 200 cos(135°) = -141.4N - F_{2y} = 200 sin(135°) = 141.4N
- Force at 225°: - F_{3x} = 300 cos(225°) = -212.1N - F_{3y} = 300 sin(225°) = -212.1N

Sum the horizontal components (ΣFx): ΣFx = 70.7 + -141.4 + -212.1 = -282.8N

Sum the vertical components (ΣFy): ΣFy = 70.7 + 141.4 + -212.1 = 0.0N

Calculate the resultant force (R): R = √((ΣFx)^2 + (ΣFy)^2) = √((-282.8)^2 + (0.0)^2) ≈ 282.8N

Example-21:

Find the equilibrium force for a system where forces of 150N, 250N, and 350N act at angles of 90°, 180°, and 270°.

Solution:

Resolve each force into its horizontal (x) and vertical (y) components:

- Force at 90°: - F_{1x} = 150 cos(90°) = 0.0N - F_{1y} = 150 sin(90°) = 150.0N
- Force at 180°: - F_{2x} = 250 cos(180°) = -250.0N - F_{2y} = 250 sin(180°) = 0.0N
- Force at 270°: - F_{3x} = 350 cos(270°) = 0.0N - F_{3y} = 350 sin(270°) = -350.0N

Sum the horizontal components (ΣFx): ΣFx = 0.0 + -250.0 + 0.0 = -250.0N

Sum the vertical components (ΣFy): ΣFy = 150.0 + 0.0 + -350.0 = -200.0N

Calculate the resultant force (R): R = $\sqrt{(\Sigma Fx)^2 + (\Sigma Fy)^2}$ = $\sqrt{(-250.0)^2 + (-200.0)^2}$ ≈ 320.2N

Example-22:

Find the equilibrium force for a system where forces of 200N, 300N, and 400N act at angles of 0°, 120°, and 240°.

Solution:

Resolve each force into its horizontal (x) and vertical (y) components:

- Force at 0°: - F_{1x} = 200 cos(0°) = 200.0N - F_{1y} = 200 sin(0°) = 0.0N
- Force at 120°: - F_{2x} = 300 cos(120°) = -150.0N - F_{2y} = 300 sin(120°) = 259.8N
- Force at 240°: - F_{3x} = 400 cos(240°) = -200.0N - F_{3y} = 400 sin(240°) = -346.4N

Sum the horizontal components (ΣFx): ΣFx = 200.0 + -150.0 + -200.0 = -150.0N

Sum the vertical components (ΣFy): ΣFy = 0.0 + 259.8 + -346.4 = -86.6N

Calculate the resultant force (R): R = √((ΣFx)^2 + (ΣFy)^2) = √((-150.0)^2 + (-86.6)^2) ≈ 173.2N

Example-23:

Find the equilibrium force for a system where forces of 180N, 280N, and 380N act at angles of 60°, 150°, and 270°.

Solution:

Resolve each force into its horizontal (x) and vertical (y) components:

- Force at 60°: - F_{1x} = 180 cos(60°) = 155.9N - F_{1y} = 180 sin(60°) = 90.0N
- Force at 150°: - F_{2x} = 280 cos(150°) = -242.5N - F_{2y} = 280 sin(150°) = 140.0N
- Force at 270°: - F_{3x} = 380 cos(270°) = 0.0N - F_{3y} = 380 sin(270°) = -380.0N

Sum the horizontal components (ΣFx): ΣFx = 155.9 + -242.5 + 0.0 = -86.6N

Sum the vertical components (ΣFy): ΣFy = 90.0 + 140.0 + -380.0 = -150.0N

Calculate the resultant force (R): R = √((ΣFx)^2 + (ΣFy)^2) = √((-86.6)^2 + (-150.0)^2) ≈ 173.2N

Example-24:

Find the equilibrium force for a system where forces of 220N, 320N, and 420N act at angles of 45°, 135°, and 225°.

Solution:

Resolve each force into its horizontal (x) and vertical (y) components:

- Force at 45°: $F_{1x} = 220 \cos(45°) = 155.5N$ $F_{1y} = 220 \sin(45°) = 155.5N$
- Force at 135°: $F_{2x} = 320 \cos(135°) = -226.2N$ $F_{2y} = 320 \sin(135°) = 226.2N$
- Force at 225°: $F_{3x} = 420 \cos(225°) = -296.9N$ $F_{3y} = 420 \sin(225°) = -296.9N$

Sum the horizontal components (ΣFx): $\Sigma Fx = 155.5 + -226.2 + -296.9 = -367.6N$

Sum the vertical components (ΣFy): $\Sigma Fy = 155.5 + 226.2 + -296.9 = 84.8N$

Calculate the resultant force (R): $R = \sqrt{(\Sigma Fx)^2 + (\Sigma Fy)^2} = \sqrt{(-367.6)^2 + (84.8)^2} \approx 377.3N$

Example-25:
Find the equilibrium force for a system where forces of 260N, 360N, and 460N act at angles of 30°, 120°, and 210°.

Solution:

Resolve each force into its horizontal (x) and vertical (y) components:

- Force at 30°: $F_{1x} = 260 \cos(30°) = 225.2N$ $F_{1y} = 260 \sin(30°) = 130.0N$
- Force at 120°: $F_{2x} = 360 \cos(120°) = -180.0N$ $F_{2y} = 360 \sin(120°) = 311.8N$
- Force at 210°: $F_{3x} = 460 \cos(210°) = -398.4N$ $F_{3y} = 460 \sin(210°) = -230.0N$

Sum the horizontal components (ΣFx): $\Sigma Fx = 225.2 + -180.0 + -398.4 = -353.2N$

Sum the vertical components (ΣFy): $\Sigma Fy = 130.0 + 311.8 + -230.0 = 211.8N$

Calculate the resultant force (R): $R = \sqrt{(\Sigma Fx)^2 + (\Sigma Fy)^2} = \sqrt{(-353.2)^2 + (211.8)^2} \approx 411.8N$

Example-26:

Find the equilibrium force for a system where forces of 150N, 250N, and 350N act at angles of 0°, 90°, and 180°.

Solution:

Resolve each force into its horizontal (x) and vertical (y) components:

- Force at 0°: - $F\_\{1x\} = 150 \cos(0°) = 150.0N$ - $F\_\{1y\} = 150 \sin(0°) = 0.0N$
- Force at 90°: - $F\_\{2x\} = 250 \cos(90°) = 0.0N$ - $F\_\{2y\} = 250 \sin(90°) = 250.0N$
- Force at 180°: - $F\_\{3x\} = 350 \cos(180°) = -350.0N$ - $F\_\{3y\} = 350 \sin(180°) = 0.0N$

Sum the horizontal components (ΣFx): $\Sigma Fx = 150.0 + 0.0 + -350.0 = -200.0N$

Sum the vertical components (ΣFy): $\Sigma Fy = 0.0 + 250.0 + 0.0 = 250.0N$

Calculate the resultant force (R): $R = \sqrt{(\Sigma Fx)^2 + (\Sigma Fy)^2} = \sqrt{(-200.0)^2 + (250.0)^2} \approx 320.2N$

Example-27:

Find the equilibrium force for a system where forces of 170N, 270N, and 370N act at angles of 60°, 150°, and 240°.

Solution:

Resolve each force into its horizontal (x) and vertical (y) components:

- Force at 60°: - $F\_\{1x\} = 170 \cos(60°) = 147.2N$ - $F\_\{1y\} = 170 \sin(60°) = 85.0N$
- Force at 150°: - $F\_\{2x\} = 270 \cos(150°) = -233.8N$ - $F\_\{2y\} = 270 \sin(150°) = 135.0N$
- Force at 240°: - $F\_\{3x\} = 370 \cos(240°) = -185.0N$ - $F\_\{3y\} = 370 \sin(240°) = -320.4N$

Sum the horizontal components (ΣFx): ΣFx = 147.2 + -233.8 + -185.0 = -271.6N

Sum the vertical components (ΣFy): ΣFy = 85.0 + 135.0 + -320.4 = -100.4N

Calculate the resultant force (R): R = √((ΣFx)^2 + (ΣFy)^2) = √((-271.6)^2 + (-100.4)^2) ≈ 289.6N

1.4 Lami's Theorem

Don't worry if you don't understand the beginning. Just go through the whole section and then come back to the beginning and that should clear up any confusion you might have. Let's begin:

Lami's Theorem is a principle used in physics and engineering to solve problems involving forces acting at a point. It's especially useful when dealing with three forces in equilibrium. Here's an explanation and an example to understand.

Lami's Theorem states that if three forces acting on a point are in equilibrium, each force is proportional to the sine of the angle between the other two forces. In mathematical terms, if three forces F_A, F_B, and F_C act at a point and the angles between them are α, β, and γ respectively, then:

$F_A / \sin(α) = F_B / \sin(β) = F_C / \sin(γ)$

This means that the ratio of the magnitude of each force to the sine of the angle between the other two forces is constant.

Example:

Imagine you have a small tent tied down with three ropes. These ropes are pulled in different directions to keep the tent stable and upright. Let's say the forces in the ropes are F1, F2, and F3, and the angles between them are 120°.

1. Set Up the Problem:
- Force F1 acts at an angle of 120° to F2.
- Force F2 acts at an angle of 120° to F3.
- Force F3 acts at an angle of 120° to F1.

2. Applying Lami's Theorem:
According to Lami's Theorem:
F1 / sin(120°) = F2 / sin(120°) = F3 / sin(120°)

3. Solving for Forces:
Since sin(120°) is the same for each angle, the forces must be equal if the tent is perfectly stable.
Therefore:
F1 = F2 = F3

This means each rope is pulling with the same force to keep the tent stable.

Another Example with Different Forces

Let's make it a bit more interesting. Suppose we have a lamp hanging from the ceiling with three wires. The lamp is in equilibrium, meaning it doesn't move. The forces in the wires are:

- F_A = 30 Newtons
- F_B = 40 Newtons
- F_C = ? Newtons

The angles between the forces are:
- $\alpha = 60°$ (between F_B and F_C)
- $\beta = 90°$ (between F_A and F_C)
- $\gamma = 150°$ (between F_A and F_B)

1. Applying Lami's Theorem:
$F_A / \sin(60°) = F_B / \sin(90°) = F_C / \sin(150°)$

2. Finding F_C:
First, we need the values of the sines of the angles:
- $\sin(60°) = \sqrt{3}/2 \approx 0.866$
- $\sin(90°) = 1$
- $\sin(150°) = 1/2$

Using these values, we set up the equation:
$30 / 0.866 = 40 / 1 = F_C / 0.5$

Simplifying the left side for F_A:
$34.64 \approx 40 = 2F_C$

Solving for F_C:
F_C = 20 Newtons

This means the third wire must pull with a force of 20 Newtons to keep the lamp hanging still.

Example-28

A chandelier is hanging with the help of two cables making an equal angle with the ceiling. Calculate the tension in both cables.

Solution:

We will now draw up the free-body diagram for this. A free-body diagram is a sketch of an object (the "body") that shows all the external forces acting on it. The object is usually represented by a simple shape like a box or a dot, and the forces are represented by arrows pointing in the direction of the force. Each arrow is labeled with the type of force and its magnitude if known.

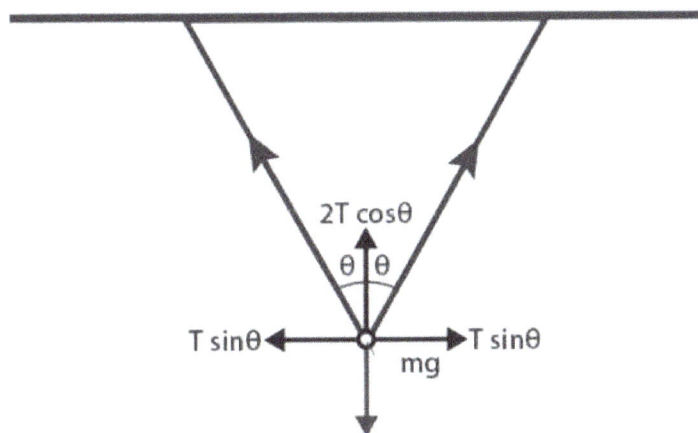

Now, we will apply Lami's theorem. But first, let's understand the FBD (Free Body Diagram). The weight of the chandelier is in the downward direction. The other force is the tension generated in both cables. Here, the tension T in both the cables is the same as the angle made by them with the chandelier is equal. (the vertical components of both the tension forces of the 2 cables are T cosθ and are working along the same line thus 2T cosθ is written in the FBD)

$$\frac{T}{\sin(180-\theta)} = \frac{T}{\sin(180-\theta)} = \frac{mg}{\sin(2\theta)}$$

since, sin (180 – θ) = sinθ and sin (2θ) = 2sinθcosθ

thus

$$\frac{T}{\sin\theta} = \frac{mg}{2\sin\theta\cos\theta}$$

Thus

T = mg / 2cosθ

Chapter-02: Moments & Couples

2.1 Moment of a force

Moment of a force = twisting effect of force about an axis of rotation.

The moment of a force is a way to describe how much a force causes something to rotate. Imagine you're trying to open a door. The force you apply to the door handle is what makes it swing open. Here's a simple way to understand it:

Definition

The moment of a force (often called torque) is the measure of the tendency of the force to cause an object to rotate around a point or axis. It's calculated by multiplying the amount of force by the distance from the point where the force is applied to the point or axis of rotation.

Formula

Moment = Force X Perpendicular distance

The above definition can be a bit difficult to understand, so let's try to understand the concept and the formula with the help of a figure.

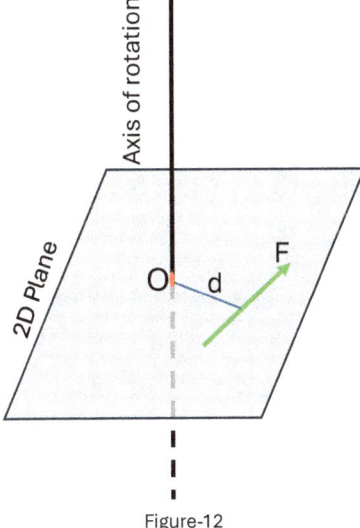

Figure-12

In figure-12

F = force

D = distance of the force from the point of rotation

O = point of rotation

The moment of the force (F) is the measure of how much torque (twisting) it can generate about the point of rotation (O). **Notice that the axis is perpendicular to the 2D plane.**

Now looking at figure-12 from above or in a technical term if we look at the top view of figure-12 then the formula for moment will become clear.

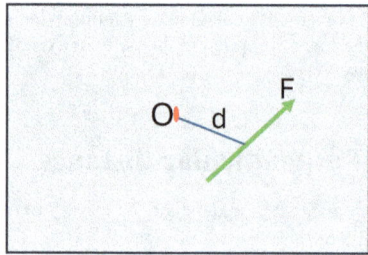

Figure-13

from Figure-13 we can tell that:

Moment = Force X Perpendicular distance between the point of rotation and line of action of the force.

Now let's try out an example of a moment to get a feel of it. If you have a door with hinges you can try this out.

Example:

Imagine you are pushing on a door to open it:

> **Force**: This is how hard you push on the door. Let's say you push with a force of 10 Newtons.
> **Distance**: This is how far from the hinges you push. Let's say you push on the door handle, which is 1 meter away from the hinges.

Using the moment formula: Moment = force times the perpendicular distance

Moment = 10N * 1m = 10N-m

But if the distance were let's say 1.2m then with the same force value we could get a higher moment:

Moment = 10N * 1.2m = 12N-m

To get a feel of what I am saying try pushing from different distances and see if it becomes easier to open the door if you increase the distance of the push from the door hinges or if you decrease it.

We can use this to our advantage since increasing one of the two gives us a different result we can get increased or decreased moment values by increasing or decreasing one of the two (force or distance) in scenarios where we won't be able to change the force value, we will try to change the distance value to get a required result similarly vice-versa.

Now, as for the unit, you can see that the unit of moment is force-distance. This distinguishes moment from work, as the unit of work is distance-force.

We have understood what a moment is and now we must understand or establish a sense of rotation. Just as for a force in chapter 1 we used the global or local axis system to determine its direction. We need to distinguish between moments that would cause:

- Clockwise rotation
- Anti-Clockwise rotation

Here, we will consider clockwise rotation as negative and clockwise rotation as positive. You can choose to do the opposite as long as you clearly state your assumptions or considerations.

To get clearer on the matter we will now do some practice problems. We will start easy for you to grasp everything then we will see systems with multiple forces.

Example-29:

Determine the moment generated by the 70N force at the center of the axis. The location of the force from the center of the axis system is given as [7, 15] meaning 7 units in the x direction from the center and 15 units in the y direction.

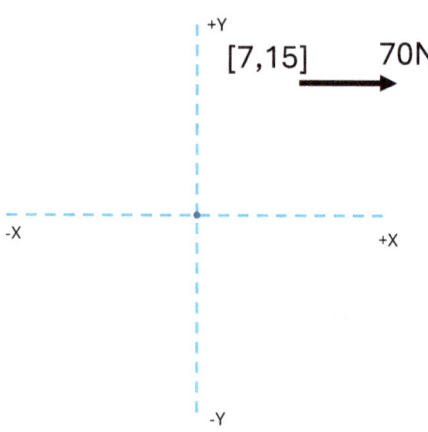

Solution:

The force is acting toward the horizontal thus the perpendicular distance would be the y-distance. As the force is facing right to generate a moment at the center it will move clockwise thus negative [imagine the lever arm (perpendicular distance just like shown in figure 12 and 13 by the distance d, to help understand the movement] Here is a link to a YT video if you are having a hard time understanding the movement: https://youtu.be/w27WGFwgs7E

Thus, Moment = force x perpendicular distance

Moment = -70 * 15 = - 1050N-unitlength

Example-30:

Do example 29 again but this time consider that the force is upward.

Solution:

Moment = 70 * 7 = 490N-unit length (to generate moment at the center it will go counter-clockwise)

Here is a figure, in case you are having a hard time imagining it.

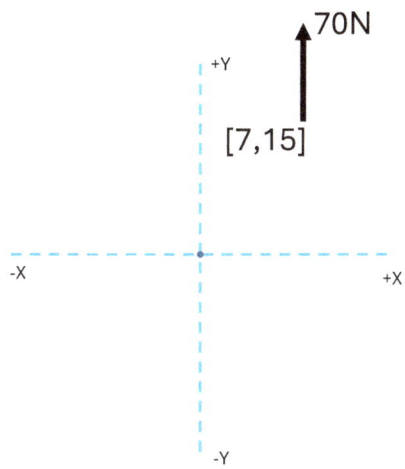

Example – 31:

Determine the moment generated by the force at the center of the axis.

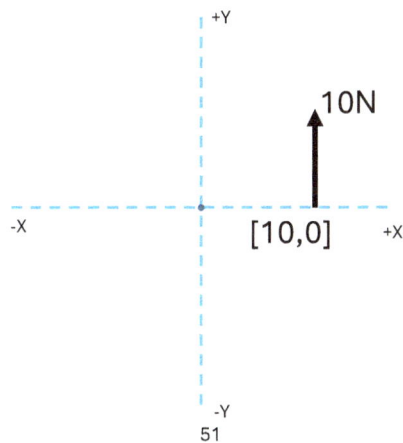

Solution:

Moment = force X perpendicular distance

Moment = 10 * 10 = 100N-unit length (Anti-clockwise thus positive)

Example – 32:

Determine the moment generated by the force at the center of the axis.

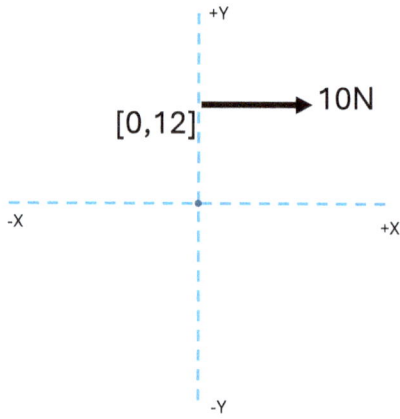

Solution:

Moment = Force X perpendicular distance

Moment = - 10 * 12 = - 120N-unit length (clockwise thus negative)

Another way of writing the answer would be Moment = 120N clockwise.

As we are using the positive and negative symbols to express whether the moment is clockwise or anti-clockwise, in your final answer you can simply write clockwise or anti-clockwise. This is a great way of documenting the answer as the sign convention you assumed (clockwise negative) might not be known to someone else but if you note down clockwise or anti-clockwise then that won't cause any confusion.

Example-33:

Determine the resultant moment at the center generated by the 2D force system.

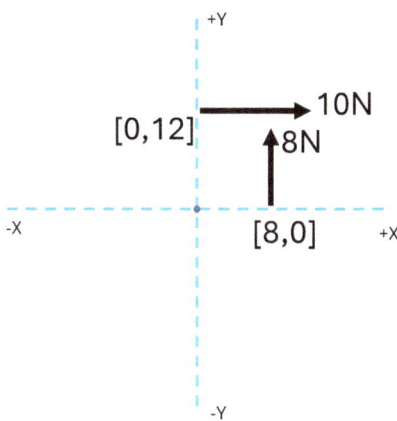

Solution:

$\sum M_O = -(10 * 12) + (8 * 8) = -56N$

Thus, the resultant moment is 56N Anti-clockwise.

Example-34:

Determine the moment generated by the force at the center of the axis.

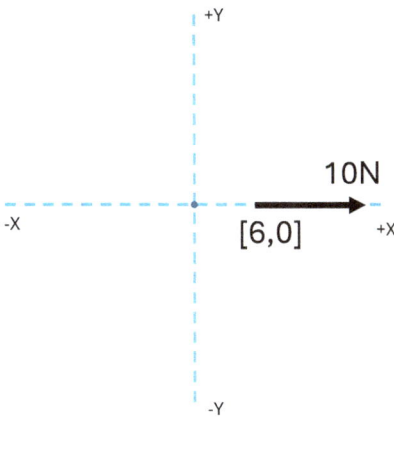

Solution:

From the formula, we know that Moment = force X perpendicular distance.

Since there is no perpendicular distance that the force makes with the center of the axis. there is also no moment generated at the center of the axis by force.

M = 10 * 0 = 0N-unitlrngth.

Example-35:

Determine the moment generated by the force at the center of the axis.

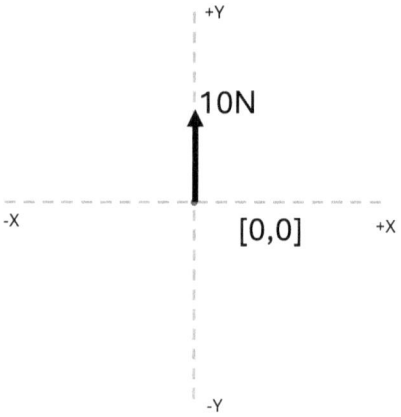

Solution:

From the formula, we know that Moment = force X perpendicular distance.

Since there is no perpendicular distance that the force makes with the center of the axis. there is also no moment generated at the center of the axis by force.

M = 10 * 0 = 0N-unitlrngth.

Example-36:

Determine the moment generated by the force at the center of the axis

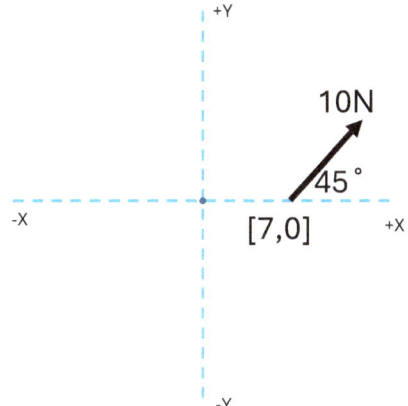

Solution:

In cases of forces that have an angle, it is better to dissolve the force into its vertical and horizontal components to carry out the moment calculation. So, here instead of one we actually have two forces:

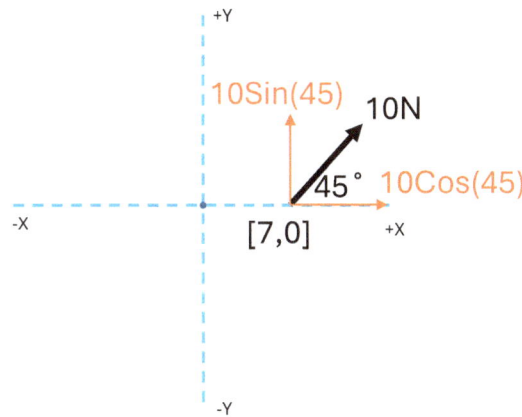

Here, the x (horizontal) component doesn't make any perpendicular distance with the center of the axis thus no moment is generated by the x component of the force.

Moment, M = (10Sin(45) * 7) – (10 Cos(45) * 0) = 49.50N-unit length

Thus, the Moment generated by the force is 49.50N in the anti-clockwise direction.

Example-37:

Determine the resultant moment generated by the 2D force system at the center of the axis.

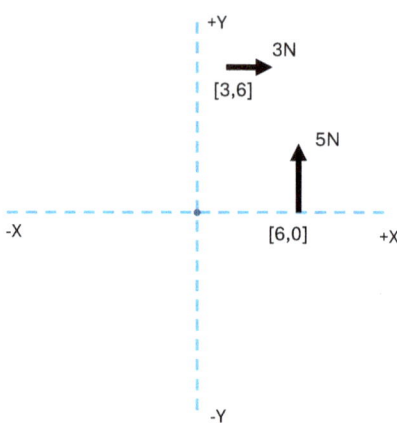

Solution:

To make it a bit easier than the calculation in example 33, we will first calculate the moments generated by the individual forces and then combine them to get the resultant.

After that we will solve it again in a more shortcut way like example 33 to get used to that way of solving as we need to develop our skill of calculating moments for simple cases in our heads and sometimes in our heads and using just a calculator for larger and massy values.

Moment generated by the 5N force is: 5 * 6 = 30N-unitlength

Moment generated by the 3N force is: -3 * 6 = -18N-unitlength

$\sum M_o$ = 30 – 18 = 12N-unitlength

Thus, the resultant moment is 12N-unit length in the anti-clockwise direction.

Now let's do it again in one line:

$\sum M_o$ = (5 * 6) – (3 * 6) = 12N-unit length

Example-38:

Determine the resultant moment generated by the 2D force system at the center of the axis.

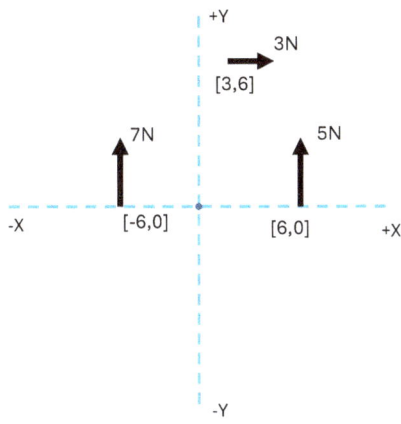

Solution:

Remember the sign convention of moments is based on rotation (clockwise or anti-clockwise) not based on the direction of the force that is generating the moment.

Moment generated by the 5N force is: 5 * 6 = 30N-unit length

Moment generated by the 3N force is: -3 * 6 = -18N-unit length

Moment generated by the 7N force is: -7 * 6 = -42N-unit length (clockwise thus negative)

$\sum M_O$ = 30 – 18 - 42 = -30N-unit length

Thus, the resultant moment is 30N-unit length in the clockwise direction.

Example-39:

Determine the resultant moment generated by the 2D force system at point P. Consider the length units to be in meters.

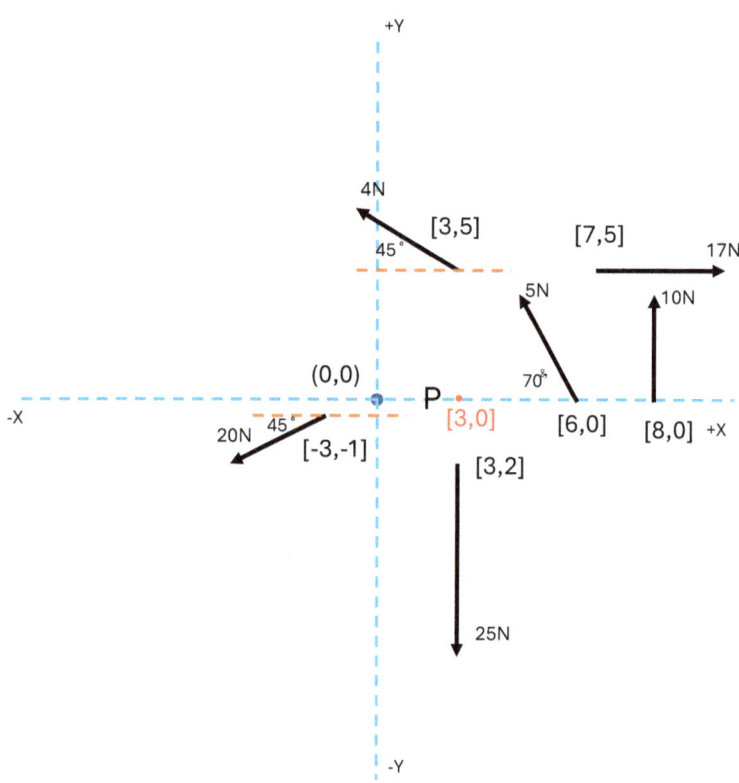

Solution:

Unlike the problems that we have done so far the point about which we will be calculating the moment is not at the center of a cartesian axis system so we have to consider that when we do our calculation.

Moment generated by 17N force: -17 * 5 = -85N-m

Moment generated by 10N force: 10 * (8-3) = 50N-m

Moment generated by 25N force: -25 * 0 = 0N-m

The moment generated by the 5N force:

>Horizontal: the horizontal component doesn't have a perpendicular distance with the point P thus no moment.

>Vertical: 5Sin(70) * (6-3) = 14.095N-m

Moment generated by the 4N force:

>Horizontal: 4Cos(45) * 5 = 14.142N-m

>Vertical: the vertical component doesn't have a perpendicular distance with the point P thus no moment.

Moment generated by the 20N force: -(20Cos(45) * 1) + (20Sin(45) * (3+3)) = 70.71N-m

$\sum M_O$ = -85 +50 +14.095 +14.142 + 70.71 = 63.947N-m (positive sign indicates the resultant moment acting on point P is anti-clockwise.

The solution in one line:

$\sum M_O$ = -(17 * 5) + (10*5) + (5Sin(70) * 3) + (4Cos(45) * 5) - (20Cos(45) * 1) + (20Sin(45) * 6)

= 63.947N-m

As moments are very important for you to master, here are a few more example problems but these problems are exactly the same except for the locations on the cartesian, so, I want you to edit these examples to make your own problems and try to solve them

Remember if you gain the skill of calculating moments quickly and flawlessly a lot of your engineering calculations will become easy for you.

If it's hard for you to imagine the schematic of the problem, then draw the problem like the previous example problems before starting to solve it

Example-40:

A force of 30 N acts upwards at the point (5, 8) on a Cartesian plane. Calculate the moment of the force about the point (1, 2).

Solution:

(5 - 1, 8 - 2) = (4, 6)
M = 30 * 4 = 120

Example-41:

A force of 40 N acts upwards at the point (6, 9) on a Cartesian plane. Calculate the moment of the force about the point (3, 4).

Solution:

(6 - 3, 9 - 4) = (3, 5)
M = 40 * 3 = 120

Example-42:

A force of 25 N acts upwards at the point (2, 6) on a Cartesian plane. Calculate the moment of the force about the point (0, 0).

Solution:

(2 - 0, 6 - 0) = (2, 6)
M = 25 * 2 = 50

Example-43:

A force of 60 N acts upwards at the point (3, 4) on a Cartesian plane. Calculate the moment of the force about the point (1, 1).

Solution:

(3 - 1, 4 - 1) = (2, 3)
M = 60 * 2 = 120

Example-44:

A force of 45 N acts upwards at the point (7, 5) on a Cartesian plane. Calculate the moment of the force about the point (3, 2).

Solution:

(7 - 3, 5 - 2) = (4, 3)
M = 45 * 4 = 180

2.2 Couple Moment

Imagine you and a friend are pushing a door open. You're both pushing with the same amount of force, but you're standing on opposite sides of the door.

- **Force**: The strength or energy you and your friend are using to push the door.
- **Direction**: You're both pushing in opposite directions; one of you is pushing from the left and the other from the right.

The Basics of a Couple Moment

A **couple** in physics means two forces that are equal in size but opposite in direction. When these forces act on an object, they create something called a **moment**. This moment makes the object rotate or turn instead of moving in a straight line.

Example:

Let's use the example of a steering wheel in a car:

1. **Forces**: When you turn a steering wheel, you usually use both hands. One hand pushes the wheel up while the other hand pushes it down.
2. **Equal and Opposite**: The force your left-hand uses to push up is the same as the force your right hand uses to push down, but they are in opposite directions.
3. **Rotation**: These equal and opposite forces don't make the steering wheel move left or right. Instead, they make the wheel rotate around its center, allowing you to steer the car.

Visualizing the Couple Moment:

Imagine you have a ruler balanced on your finger. If you push down on one end of the ruler and your friend pushes up on the other end with the same amount of force:

- The ruler won't move sideways because the forces cancel each other out.
- Instead, the ruler will rotate around your finger.

This rotation happens because the forces create a turning effect or a couple moment.

Importance:

Couple moments are important because they help us understand how things turn and rotate. Engineers and designers use this concept to create and improve tools, machines, and vehicles, ensuring they work efficiently and safely.

Keynotes:

What is a couple?

If two forces:

- Are equal in magnitude
- Are opposite in direction (i.e. parallel on a 2D plane with opposite direction)
- Don't share the same line of action.

Figure-14, shows a couple:

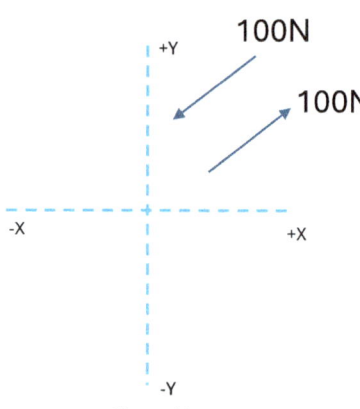

Figure-14

The moment generated by a couple is the force magnitude multiplied by the perpendicular distance between the lines of action of each force.

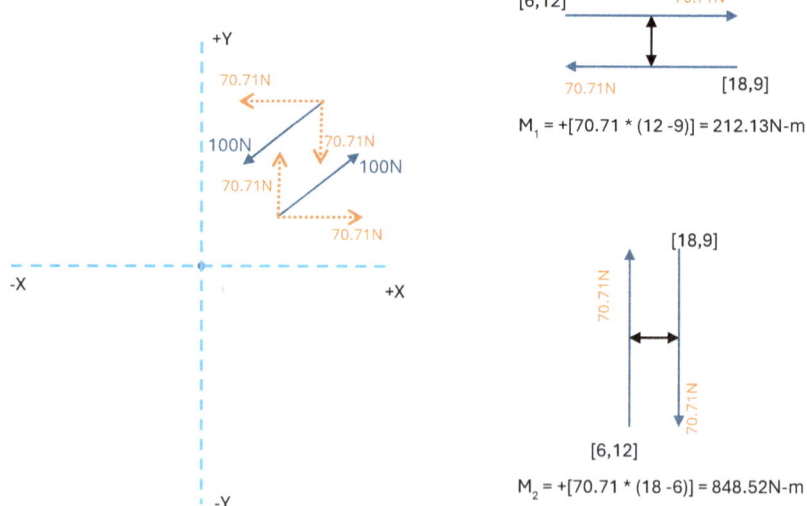

$M_1 = +[70.71 * (12 - 9)] = 212.13$ N-m

$M_2 = +[70.71 * (18 - 6)] = 848.52$ N-m

$M_{couple} = 212.13 + 848.52 = 1060.65 N\text{-}m$

The moment generated by a couple is constant, no matter where the point of rotation is located within the plane.

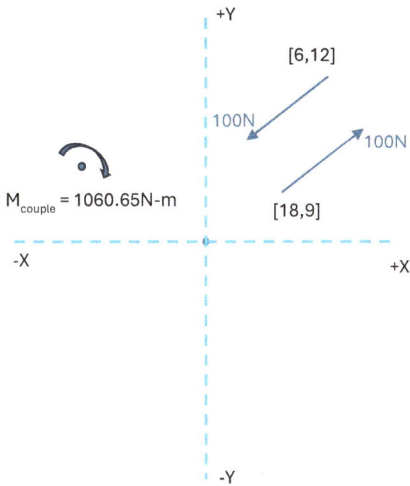

The moment generated by a couple is constant, no matter where the couple is positioned within the plane.

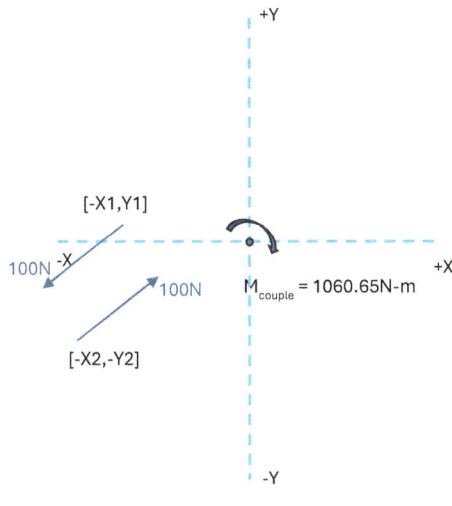

Chapter-03: Practical/Practice Problems

This chapter's purpose is to open your mind toward the application of the theories and the example problems in real-world situations. You will use nothing more than what you have already learned but you will have to learn to think outside the box to solve these problems. Sometimes you have to follow the breadcrumbs outside of the information provided in this book, as the real world and real-world scenarios are vast you should use every tool at your disposal to broaden your knowledge and application of what you have learned. As you continue solving these problems, also try to make problems from the world around you and try to solve them or see what other ways you can implement the things you have learned so far. If you have followed along while practicing all the example problems, then after solving a few of the practice problems in this chapter you'll understand what I am talking about.

To make sure you don't get frustrated I will put up the solutions of some of these practice problems on my YouTube channel.

Playlist link:

https://www.youtube.com/playlist?list=PLBdCdFKyQA-wXvQ3gkG2TbJJnCoePdAIU

PP-1

Find the length of the flagpole and the angle it makes with the cord. The length of the cord is double the distance from the pole to the ground hinge where the cord is tied.

[Hint: Brainstorm how you can measure the length of the cord if it were a real scenario]

PP-1 Simplified

Find the length of the flagpole and the angle it makes with the cord. The length of the cord is double the distance from the pole to the ground hinge where the pole is tied. Jimmy was leaning on the flagpole before he started walking toward the ground end of the cord. It took him exactly 20 steps to reach the ground end of the cord. Jimmy is a 10-year-old peculiar boy who walks a bit differently than us. He starts his next step exactly where he finished his last step (shown in the figure below).

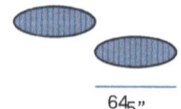

64_5"

PP-2

Sammy lives on the 9th floor of a building that is very close to an Industry. When he goes to the roof of his building, he notices that the huge conveyor belt's top is at level with the top of the building (10th-floor equivalent). He went to the industry for a school trip and learned that the conveyor belt that he sees from his home is 50 meters in length. What is the angle between the conveyor and the ground?

[Hint: what is the typical story height (approximate) of a residential building?]

The following are some images representing a similar setup of the conveyor belt. For those who are not familiar with it. It will help imagine the scenario.

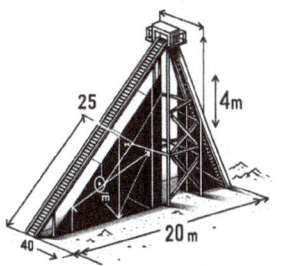
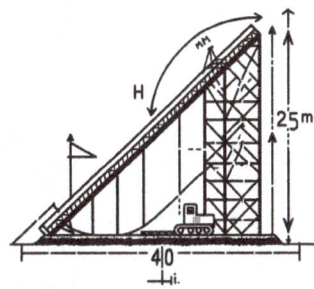

PP-3

Calculate the length of the pole. Consider the pole to be a perfectly vertically straight pole. (it's an assumption you can use).

[Hint: I'm 5'11" tall. Yes, that's me in the picture. Just like PP-1, it takes me 8.5 steps from the pole to the ground end of the cable. And each step is approximately 1' (a little over 1' actually as shown in the pictures but for ease of calculation you can consider 1')]

Sorry about the worn-out shoes. I really like this shoe, so haven't moved on to a new shoe yet.

PP-4

In the following figure, the weight acting as a nail underground has a mass of 50kg. the boy has a pull force of 150N as measured from a physical competition he took part in. The boy also took part in a rope-pulling game where his opponent was the cow. He lost to the cow and exclaimed that he felt the cow was twice as strong as him. Now as you can see both of them are pulling the nail. So, will the nail come out at the moment (captured by the figure below) they apply their maximum force? There are many factors in play here in reality but answer it only based on your knowledge of 2D concurrent coplanar forces. First, try to answer without calculating (guess the trajectory of the resultant force), then do calculations to come to a conclusion and also to see if your educated guess was right or not.

[Hint: the reflection of examples 6 and 7 from Chapter 2, (1kg X 9.81m/s^2 = 9.81N)]

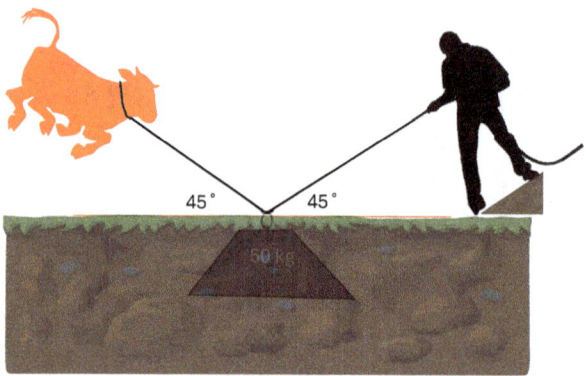

PP-5

In a circus, 4 people pull a ring onto place at center stage and a pigeon goes through the ring while the ring is on fire, thus it is necessary to have the ring in the desired location for the performance as well as the safety of the bird. The 4 people are always the same people and they have figured out how strongly to pull the ring to keep it in place and coincidentally it requires them to pull with their full force. However, it's actually not a coincidence as an engineer came up with the angle of the pull for each man considering their full force capacity. And the circus doesn't resort to the ring being tied to some fixture instead of the men pulling it in place for aesthetics, Afterall it's a circus. Now, out of the 4 men, 1 fell ill and sent his brother to work in his stead, however his brother's pull force (40N) is different from him, and the circus has enough money to hire one more person to maintain the equilibrium. The circus has the means to measure someone's pull force. You are appointed as the trainee engineer to recruit the 5th person and to place him where necessary to keep the ring where it should be. Use the following figure for reference.

[Hint: the line of action for the ropes pulling the ring and the angles are found from the original engineer's schematics given below. The schematic displays the forces exerted by the current 4 people on the day the original 4th got sick, remember example 12 from chapter 1?]

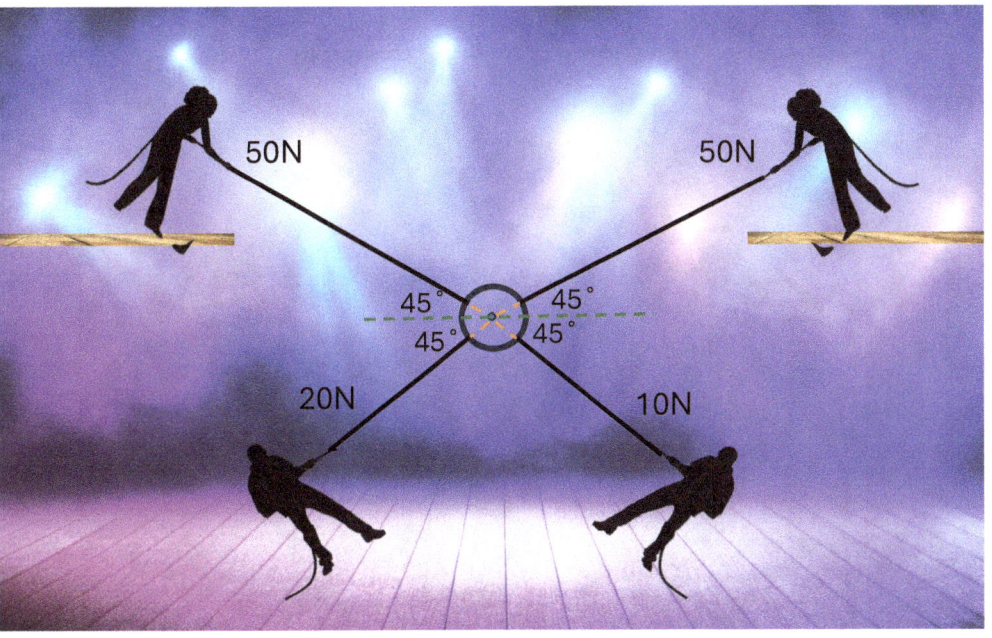

PP-6

A young engineer bought a house, which was previously owned by his boss. His boss had made a swing using the trees in the backyard, a spare tire, and some cables. The tensile strength capacity of the cables is known (850N) to the young engineer as those are the cables they seldom use for the projects that they do. The weight of the tire is also known to him (20kg) as it is a very common tire. He is a father of one, and his child weighs about 50kg. The engineer made the calculations as to whether the DIY swing was safe for his child or not. What was his decision? Did he deem it safe?

[Hint: in the image, some information is given that might be beneficial to you]

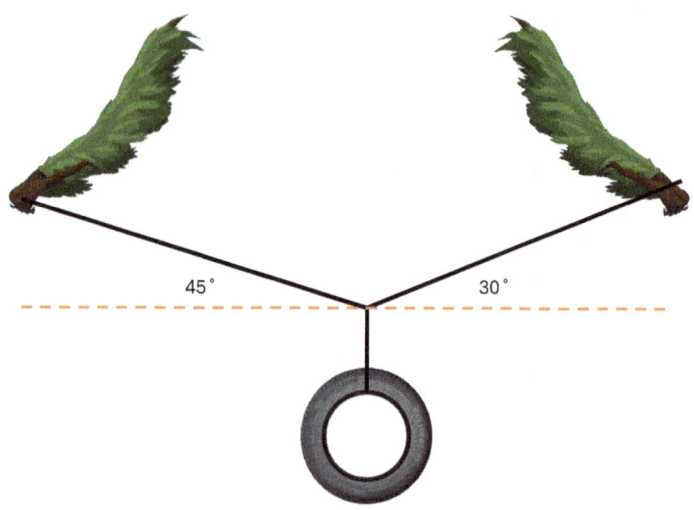

PP-7

Timmy and Bobby, two brothers, are helping their dad pull his broken-down car inside the garage. The two boys tied the car with two ropes but since they had had a fight earlier, they were not getting close to pulling the car they were staying on the opposite side of the road. Their dad who is an engineer sees this from the 2nd floor of their house and makes a comment. As their father and as a very perceptive individual he can guess their pull strength quite well. Timmy- 250N, Bobby- 200N. As no one is in the driver's seat of the car they need to pull the car straight. From the schematic of the situation from the father's eyes given below what was the comment that he made?

- A) You guys are doing a good job
- B) Don't fight with each other
- C) Wait the car will deviate from being straight slightly

[Hint: if you have done all the example problems from the 1st chapter and practiced a lot too then you can answer this without even calculating, well if you have good deductive skills you can answer it without calculating as well, just by analyzing my storytelling pattern but the purpose of this question is not for putting your deductive skills to the test but rather your skills in understanding how concurrent coplanar forces work.]

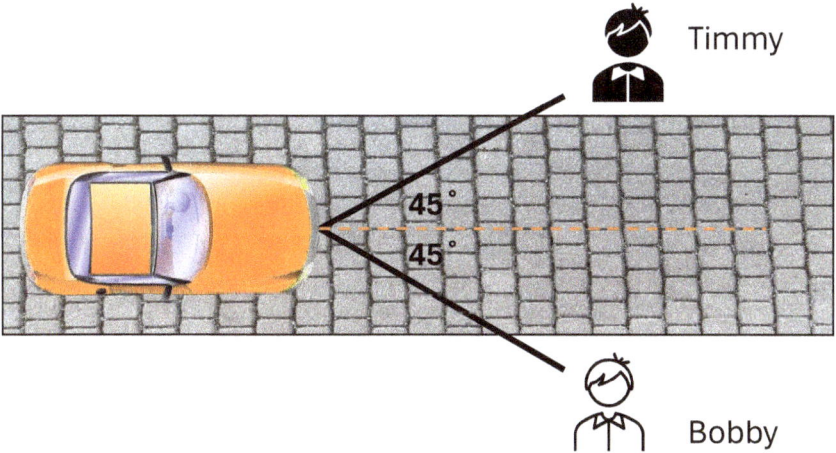

PP-8

From the pictures below you can see a light hanging from the ceiling with three supports. Each of the supports consists of 3 wires. If the supports were equally placed, then calculate the force on the wires.

[Hint: This is a bit of an advanced problem if you are not used to solving problems that provide you with almost all the information you need. But in real work no one will provide you with all the information, that's why we have surveys and educated assumptions and whatnot. At your disposal, you have some tools too. You have the pictures to determine what type of light it is. I'd do a reverse image search on Google and try to find out the weight of the light then divide it in three and so on. You are welcome to watch my take on the solution to this question but remember that my approach may differ from yours.]

[**Purpose:** the purpose of this question is to open your mind towards the everyday things that surround you from an engineer's POV (point of view) so that you can make more real-life questions like this and try to solve them to increase your skill.]

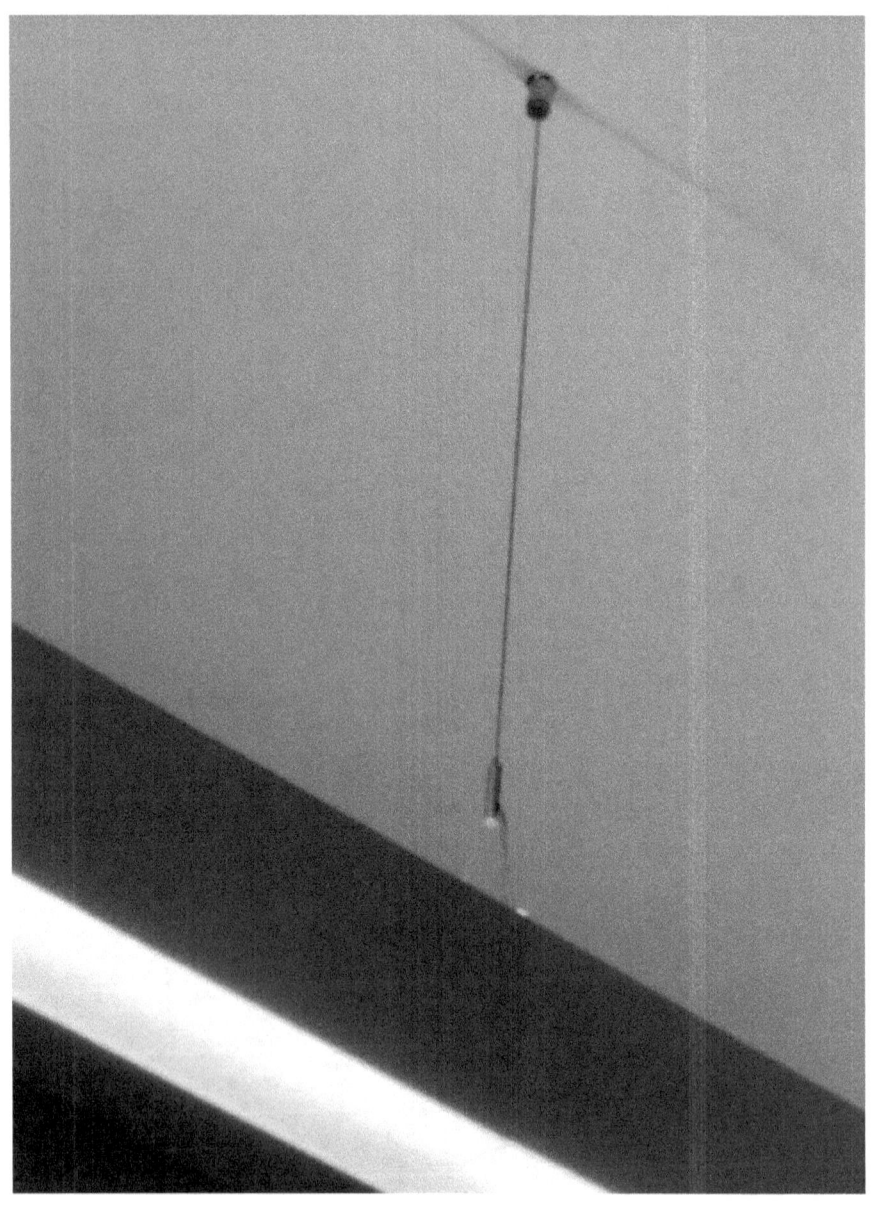

The photos are from a classroom in the Myhall building at the University of Toronto.

Zahed Zisan's Sample Exam

Department of Civil Engineering

Course Title: Applied Mechanics Part 1

Time: 1 Hour   Full Marks: 0.00

---

1. Calculate the resultant and the angle it makes with the horizontal for the force system shown in Figure - 1. [Marks: 0]

2. If all the forces in figure-1 were 30N then what would be the resultant and the angle with the horizontal? Can you answer without calculating? [Marks: 0]

3. Calculate the resultant and the angle of the resultant with the horizontal for the forces in Figure - 2 and can you guess if the resultant would be closer to the Y or the X axis before you do the calculations? [Marks: 0]

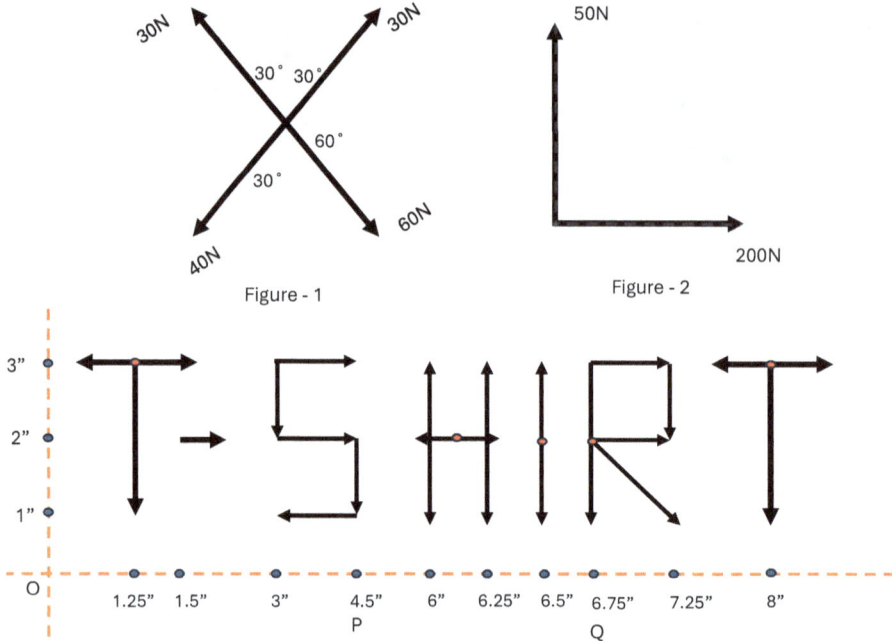

Figure - 1

Figure - 2

Figure - 3

4. If all the forces in Figure – 3 have a value of 35lb, what would be the resultant moment about point O? [Marks: 0]

5. If all the forces in Figure – 3 have a value of 40lb, what would be the resultant moment of the horizontal forces about point P? [Marks: 0]

6. If all the forces in Figure – 3 have a value of 40lb, what would be the resultant moment of the vertical forces about point P? [Marks: 0]

Note: For question 4,5,6, consider that the diagonal force makes an angle of 45° with both the horizontal and the vertical.

Story:

I used to wear Shirts every day, but after starting grad school at U of T, to save time and also the hassle of ironing shorts I have switched to T-Shirts! And T-shirts can take a lot of moment/torque compared to Shirts. And even if T-shirts are wrinkly, it doesn't matter that much compared to if it were shirts. So, my conclusion was moments can go wild in the case of a T-shirt and that's the inspiration behind the T-SHIRT being the moment question.

Appendix

Sine, Cosine, and Tangent in a Right-Angled Triangle

In a right-angled triangle, there are three important trigonometric ratios: sine (sin), cosine (cos), and tangent (tan). These ratios help us relate the angles of a triangle to the lengths of its sides.

The Sides of a Right-Angled Triangle

1. Hypotenuse: The longest side, opposite the right angle.

2. Opposite: The side opposite the angle you are working with.

3. Adjacent: The side next to the angle you are working with, but not the hypotenuse.

Sine (sin)

The sine of an angle is the ratio of the length of the opposite side to the hypotenuse.

$\sin(\theta)$ = Opposite / Hypotenuse

Cosine (cos)

The cosine of an angle is the ratio of the length of the adjacent side to the hypotenuse.

$\cos(\theta)$ = Adjacent / Hypotenuse

Tangent (tan)

The tangent of an angle is the ratio of the length of the opposite side to the adjacent side.

$\tan(\theta)$ = Opposite / Adjacent

Example

Let's say we have a right-angled triangle with one of the angles being θ.

1. The hypotenuse is 10 units long.

2. The opposite side to angle θ is 6 units long.

3. The adjacent side to angle θ is 8 units long.

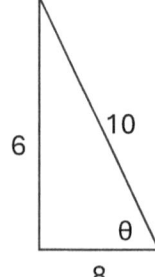

For this triangle:

- $\sin(\theta) = 6 / 10 = 0.6$

- $\cos(\theta) = 8 / 10 = 0.8$

- $\tan(\theta) = 6 / 8 = 0.75$

These ratios help us understand the relationships between the angles and sides of the triangle.

Appendix Problems

Problem-1

Find A

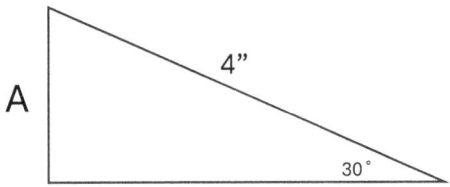

Problem-2

Find A

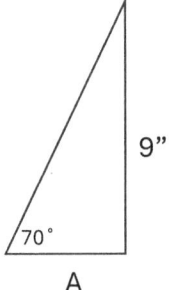

Problem-3

Find A

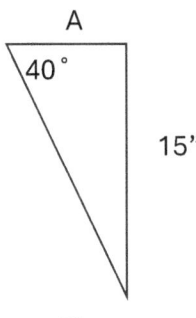

Problem-4

Find A

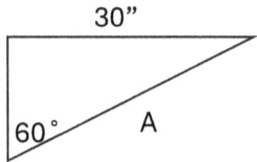

Problem-5

Find A & B

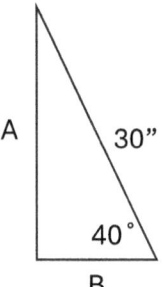

Problem-6

Find A

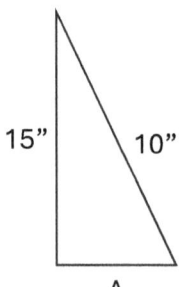

Problem-7

Find A & B

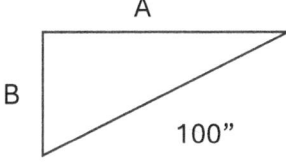

Problem-8

Find A & B

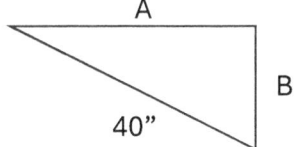

Solutions:

Problem-1

Find A

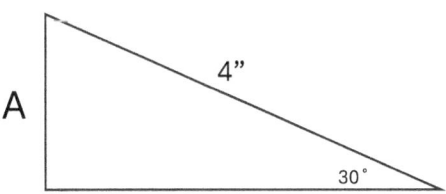

$Sin 30° = \frac{A}{4"}$, thus A = 2" [A = 4" $Sin 30°$ = 2]

Problem-2

Find A

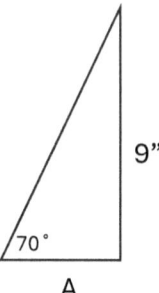

$\cos 70° = \frac{A}{9"}$, thus A = 3.07"

Problem-3

Find A

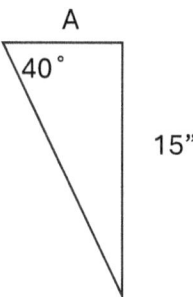

$\tan 40° = \frac{15"}{A}$, thus A = 17.87"

Problem-4

Find A

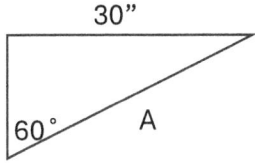

$\sin 60° = \frac{30"}{A}$, thus A = 23.09"

Problem-5

Find A & B

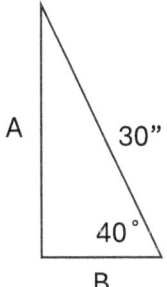

$\sin 40° = \frac{A}{30"}$, thus A = 19.28"

$30^2 = 19.28^2 + B^2$

Thus, B = $\sqrt{30^2 - 19.28^2}$ = 22.98" or approximately 23"

Problem-6

Find A

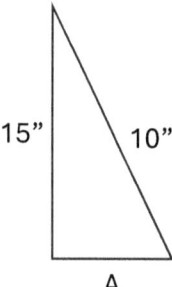

$15^2 = \sqrt{A^2 + 10^2}^2$

Thus, $A = \sqrt{15^2 - 10^2} = 11.18"$

Problem-7

Find A & B

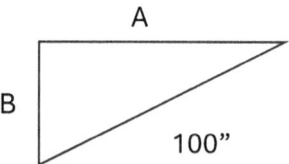

$Sin45° = \frac{B}{100"}$, thus B = 100" Sin45° = 70.71"

$Cos45° = \frac{A}{100"}$, thus A = 70.71"

Problem-8

Find A & B

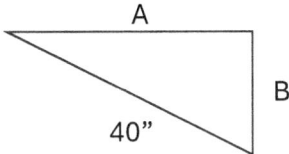

$Sin 30° = \frac{B}{40"}$, thus B = 40" Sin30° = 20"

$Cos 30° = \frac{A}{40"}$, thus A = 34.64"

www.ingramcontent.com/pod-product-compliance
Lightning Source LLC
Chambersburg PA
CBHW071949210526
45479CB00003B/860